Practical Paranoia™
iOS 16
Security Essentials

- ☑ **The Easiest**
- ☑ **Step-By-Step**
- ☑ **Most Comprehensive**
- ☑ **Guide To Securing Data and Communications**
- ☑ **On Your Home and Office iPhone and iPad**

Marc L. Mintz, MBA-IT, ACTC, ACSP

TPP
The Practical Paranoid™

Highest Praise for
Practical Paranoia Security Essentials

★ ★ ★ ★ ★

The author claims this is for the "new to average user (not the IT Professional) in mind." I'm not convinced of that. I think there are many IT Professionals who could learn a great deal from this book. That comment is not meant to denigrate IT Professionals (I are one) but to indicate just how much valuable information is included in this book. And for the beginner or average user who continually read about the vulnerability of even the Macintosh community "hooked" to the internet, this book lays it out how to protect yourself from those less than "slugs" who want nothing better than to empty your bank account.

- Santa Fe Cowboy (Amazon Verified Purchase Review)

★ ★ ★ ★ ★

My wife's e-mail got hacked about ten days ago, and that provided the impetus to finally tackle security issues for my Mac. I know I've been living on borrowed time.

Practical Paranoia arrived two days ago, and I started to glance through it at lunch yesterday. Well, three hours later I had "glanced" through most of it. It is an easy read.

This book is exactly what I hoped for... a little bit of data to convince me one last time that I need to do this, a little anecdotal humor, and then explicit step by step instructions (screen shots) on everything I need to do to protect my computer and my data. Marc even provides "assignments" to make sure the "reader "gets it."

Not only are we instructed on ways to protect us from the evil outside, but there are also instructions to protect data from hardware malfunctions.

This is exactly the way I need to learn and do.

- Photodog (Amazon Verified Purchase)

★ ★ ★ ★ ★

I became a client of Marc's five or six years ago. So, when Marc announced his new book, Practical Paranoia, I didn't waste any time buying a copy.

The reason is simple, when Marc tells you something, you can take it to the bank.

The book is clean, well-organized, designed to be completely accessible for the non-geek, and provides step-by-step screen shots and instructions on how to save yourself from the nightmare of data loss or breach.

I can tell you from my own bitter experience with a security breach we experienced on the cloud a few years ago (not Marc's purview) that cost me more than two months of my life, an ounce of prevention is worth a *ton* of cure.

And in this book, for a few bucks, Marc has laid out the preventative roadmap that can save you more time, money, and misery than you may imagine.

Former Intel CEO Andy Grove entitled his memoirs Only the Paranoid Survive. In the wild world of connected everything, Practical Paranoia is an important guide to survival.

Buy it. Implement it. You'll sleep better.

- Lanny Goodman (Amazon Verified Purchase)

★ ★ ★ ★ ★

Easy to use - I did about half of the recommendations in under an hour. I'm a writer and I run a business from home. I'm not so technically adept. When I travel, I hate packing up all my equipment and putting things in storage to make sure that, in the unlikely event the house is robbed, all my data is safe, and I could easily restart my business from a backup in the cloud. This book is a "paint-by-numbers" approach to security for every aspect of a computer system, from backups to email to encrypting the physical drives. There's really nothing else on the subject unless you want to read technical blogs for days. This book has it all in one place and makes it easy for even a normal person (like me).

- Ben Taxy (Amazon Verified Purchase)

Copyright

Practical Paranoia: iOS 16 Security Essentials

Author: Marc Mintz

Copyright © 2022 and 2023 by The Practical Paranoid, LLC.

Edition: v6 20230111 (iOS 16.2 and iPad OS 16.2)

Cover design by Ed Brandt
Paperback ISBN: 978-1-949602-03-6

Dedication

To Candace,
without whose support and encouragement
this work would not be possible

Contents at a Glance

Contents in Detail

PRACTICAL PARANOIA
iOS 16
SECURITY ESSENTIALS

MARC L. MINTZ, MBA-IT, ACTC, ACSP

Proof of Purchase

Practical Paranoia iOS 16 Security Essentials

Full Name: _____

Mailing Address: _____

_____

_____

Email Address: _____

Phone: _____

Educational Institution: _____

Position: _____

For what are you redeeming your Proof of Purchase? _____

_____

_____

Please attach a copy of your receipt

Mail to:

The Practical Paranoid, LLC
1000 Cordova Pl #842
Santa Fe, NM 87505

Thank You for Studying Practical Paranoia!

Dear Student,

Thank you for getting this far into this book. Although I cannot promise it will be as easy getting all the way through as it was to get here, I do promise this is the easiest and most comprehensive book in this category that you can buy.

When I wrote the first edition of Practical Paranoia, I received many emails and calls from instructors, students, and fans thanking me for the book. In truth, over half of this book came out of the questions and insights provided by the readers themselves. I love the feedback. I invite you to write to me at *marc@thepracticalparanoid.com*.

I also ask a favor. Please write a review of *Practical Paranoia*. Loved it, hated it, what worked for you, what you would like to see added or changed. I both enjoy and value your feedback.

Reviews can be difficult to come by these days. You, the reader, have the power now to make, break, and shape the evolution of a book. If you have the time, please visit my author page on Amazon.com[1]. Here you can find all my books and leave a review.

Thank you so much for studying Practical Paranoia and for spending time with me.

Warmly,

[1] *https://www.amazon.com/author/marclmintz*

1 Introduction

Just because you're paranoid doesn't mean they aren't after you.
–Joseph Heller[1], Catch-22

Everything in life is easy–once you know the how.
–Marc L. Mintz[2]

What You Will Learn in This Chapter

- What is Cybersecurity and Internet Privacy
- Who should study this course
- What is unique about this course
- Why worry?
- Reality check
- About the author
- Practical Paranoia updates
- Notes for instructors, teachers, and professors
- Update bounty
- Format conventions used in this book
- What is New in iOS 16 cybersecurity and internet privacy

---

[1] *https://en.wikipedia.org/wiki/Joseph_Heller*
[2] *https://thepracticalparanoid.com*

What You Will Need in This Chapter

- No additional resources required.

1.1   What is Cybersecurity and Internet Privacy

Cybersecurity

Merriam-Webster dictionary defines cybersecurity:

> measures taken to protect a computer or computer system (as on the Internet) against unauthorized access or attack.

I think the *Cybersecurity & Infrastructure Security Agency (CISA)*[3] says it better:

> Cybersecurity is the art of protecting networks, devices, and data from unauthorized access or criminal use and the practice of ensuring confidentiality, integrity, and availability of information.

Internet Privacy

Wikipedia defines internet privacy:[4]

> involves the right or mandate of personal privacy concerning the storing, repurposing, provision to third parties, and displaying of information pertaining to oneself via Internet.

Taken at the most fundamental level, cybersecurity and internet privacy are about protecting your personal information (passwords, sites visited, health conditions, finances, phone calls, emails, text messages, relationships, private photographs, video and audio recordings, public and private life, and so much more) from unauthorized access and use.

---

[3] *https://us-cert.cisa.gov/ncas/tips/ST04-001*
[4] *https://en.wikipedia.org/wiki/Internet_privacy*

As our computers, mobile devices, and internet services are the central hub for storing and transmitting all our data, it makes sense to focus our resources on securing the Information Technology (IT) in our lives.

1.2   Who Should Study This Course

Traditional business thinking holds that products should be tailored to a laser-cut market segment. Something like *18-25-year-old males, still living at their parents' home, who like to play video games, working a minimum-wage job*. Yup, we all have a pretty clear image of that market segment.

In the case of this security and privacy course, the target market segment is *all users of iPads and iPhones*. Really! From my great-Aunt Rose who is wrestling with using her first device, to the small business, to the Information Technology (IT) staff for major corporations and government agencies: this is for all of you.

There is little difference between *home-level security* and *military-grade security* when it comes to cybersecurity and privacy technology. Even though the military may use better security on their physical front doors (e.g., MPs with machine guns protecting the underground bunker compared to a home with a Kwikset deadbolt and a neurotic Chihuahua) the steps to secure computers and mobile devices for home and business use are almost identical for both private and military users.

The importance of data held in your device may be every bit as important as the data held by the CEO of a Fortune 500. Your data also is as vulnerable to penetration as the data of any other person, and the repercussions of that data being accessed are just as important. How would it feel if you lost passwords, financial information, travel scheduling, personal relationship discussions and photos of loved ones?

1.3   What is Unique About This Course and Book

By following the easy, illustrated, step-by-step instructions in this book, you can secure your computer to industry standards. The steps outlined here are the same steps used by my consulting organization when securing systems for hospitals, government agencies, and the military.

Practical Paranoia Security Essentials is the first comprehensive security book series written with the new to average user in mind as well as IT professionals and STEM and Computer Science students. Hardening your cybersecurity and privacy helps your business protect the valuable information of you and your customers.

Should your work include HIPAA, SEC, or legal-related information, to be in full compliance with regulations it is likely that you need to be using the latest operating system version. Newer operating system (OS) versions have newer protections and filters.

Do not let the number of pages here threaten you. This book is the ultimate step-by-step guide for protecting your device, but dull background theory is reduced to a minimum. I include only what is necessary to grasp the need-for and how-to. All information and steps are built on current guidelines, policies and procedures, and best practices from Apple, Google, Microsoft, NIST, NSA, US-CERT, and my own 34 years as an IT consultant, developer, technician, and trainer.

The organization of this book is simple. Each chapter represents a major area of security and privacy vulnerability, along with the tasks you should do to protect your data, device, and personal identity. To review your work using this guide, use the *Security Checklist* provided at the end of this book.

To make your system as secure as possible, I recommend following the sequence provided in this guide. Bad guys seek out your weak points. Leave no obvious weakness and they move on to an easier target. There is an old joke: Two friends are camping in the forest when they see a bear charging toward them. The first friend asks the second: *How fast do you need to run to out-run a bear*? The second friend replies: *Just a little faster than you.*

Theodore Sturgeon, American science fiction author and critic, stated: *Ninety percent of everything is crap*[5]. Mintz's extrapolation of Sturgeon's Revelation is *Ninety percent of everything you have learned and think to be true is crap.*

Sturgeon was an optimist.

---

[5] *https://en.wikipedia.org/wiki/Sturgeon%27s_law*

I have spent most of my adult life in distilling what is real and accurate from, well, Sturgeon's 90%. The organizations and workshops I have produced, and the *Practical Paranoia* book series, all spring from this pursuit. If you find any area of this workshop or book that you think should be added, expanded, improved, or changed, contact me personally with your recommendations at *marc@thepracticalparanoid.com.*

1.4   Why Worry?

In terms of network, Internet, and data security, all users of computers and mobile devices must be vigilant because of the presence of malware[6], physical theft, malicious websites, and cybercriminals. Attacks on computer and mobile device users by tricksters, criminals, and governments are on a steep rise. How bad is the situation?

- Per a study from Ponemon Institute and IBM Security, the global average cost of a data breach in 2019 was $3.92 million.

- According to data from Kensington and the FBI, one laptop is stolen every 53 seconds, and over 70 million cell phones are **lost** each year. Of those millions of devices stolen or lost, only 3% ever are recovered.

- Typical email is clearly readable at dozens of points along the Internet highway on its trip to the recipient and may be read by someone you do not know.

- If you are in the USA, your government monitors your devices. The Cyber Intelligence Sharing and Protection Act (CISPA)[7] allows the government easy access to all your electronic communications. PRISM[8] allows government agencies to collect and track data on any American device. Do not feel smug if you're not in the USA, because you're likely tracked at least as vigorously.

---

[6] *https://en.wikipedia.org/wiki/Malware*
[7] *https://en.wikipedia.org/wiki/Cyber_Intelligence_Sharing_and_Protection_Act*
[8] *https://en.wikipedia.org/wiki/PRISM_(surveillance_program)*

The list goes on, but you get the point. It is not a matter of *if* your data will ever be threatened. It is only a matter of *when,* and how often attempts will be made.

1.5   Reality Check

Nothing can 100% guarantee 100% security 100% of the time. Even the White House and CIA websites and internal networks have been penetrated. We know that organized crime, as well as the governments of China, North Korea, Russia, Great Britain, United States, Australia, and other nations have billions of dollars and tens of thousands of highly skilled security personnel on staff looking for *zero-day exploits*[9]. The zero-day exploits are vulnerabilities not yet discovered by the developer. As if this is not enough, the U.S. government influences the development and certification of most security protocols. This means that industry-standard tools used to secure our data have been found to include vulnerabilities introduced by U.S. government agencies.

With these odds against us, should we just throw up our hands and accept there is no way to ensure our privacy? Well, just because breaking into a locked home only requires a rock through a window, should we give up and not lock our doors? Of course not. We do everything we can to protect our valuables and not become victims. When leaving on vacation we lock doors, turn on motion detectors, notify the police to prompt additional patrols, and stop mail and newspaper delivery.

The same is true of our digital lives. For the very few who are targeted by the NSA, FBI, CIA, etc., there is little that can be done to completely block them from reading your email, following your chats, and recording your web browsing. But you can make it time and labor intensive for them to do so.

And you *can* protect yourself, your data, and your devices from being penetrated by criminals, pranksters, competitors, nosy people, a wackadoodle ex, as well as about the collateral damage caused by malware. By following this book, you can fully secure your data and your first device in two days, and any additional devices in a half day. This is a small price to pay for peace of mind and security.

---

[9] *https://en.wikipedia.org/wiki/Zero-day_(computing)*

It is imperative that you secure all points of vulnerability. Remember, penetration usually does not occur at your strong points. A home burglar avoids hacking at a steel door when a simple rock through a window gains entry. A strong password and encrypted storage by themselves do not protect data (including usernames and passwords) from hackers.

- Note: Throughout this book we provide suggestions on how to use various free and for-fee applications to help enforce your protection. We have used these applications with success, and thus feel confident in recommending them. Neither Marc L. Mintz, nor The Practical Paranoid, LLC. receive compensation for suggesting applications and we may change suggestions as our experience changes.

1.6   About the Author

Marc Louis Mintz is a respected IT consultant and trainer with over three decades of experience. Marc's enthusiasm, humor, and training expertise were honed on leading edge work in motivation, management development, and technology. His software and hardware workshops are highly rated by seminar providers, meeting planners, managers, and participants because he empowers participants to see with new eyes, think in a new light, and problem solve using new strategies. Marc holds a Master of Business Administration with specialization in IT (MBA-IT), Chauncy Technical Trainer certification, Post-Secondary Education credentials, and over a dozen industry certifications.

When away from the podium, Marc is in the trenches, working to keep his clients' systems secure and private. Well, not so much anymore. As of 2020 Marc retired as CEO and CIO of Mintz InfoTech, Inc. to spend more time with his passions: the world's best partner and dogs, Voice Over Acting, and maintenance of the very best cybersecurity and internet privacy guides.

The author may be reached at:

Marc L. Mintz
The Practical Paranoid LLC.
1000 Cordova Pl #842
Santa Fe, NM 87505

Email: *marc@thepracticalparanoid.com*
Web: *https://thepracticalparanoid.com*

1.7   Practical Paranoia Updates

Information regarding IT security changes daily, so we offer many options to keep you on top of everything.

1.7.1   Practical Paranoia Paperback Book Updates and Upgrades

We are constantly updating *Practical Paranoia* so that you have the latest, most accurate resource available. If at any time you wish to upgrade to the latest version of *Practical Paranoia* at the lowest price we can offer:

1. Tear off the front cover of your ***Practical Paranoia***.

2. Make your check payable to *The Practical Paranoid* for $32.50.

3. Send front cover, check, and mailing information to:
    The Practical Paranoid, LLC
    1000 Cordova Pl #842
    Santa Fe, NM 87505

4. Your new copy of Practical Paranoia is sent by USPS. Please allow up to four weeks for delivery.

1.7.2   Practical Paranoia Kindle Updates and Upgrades

We are constantly updating *Practical Paranoia* so that you have the latest, most accurate resource available. Unfortunately, we have no control over Kindle version upgrades (such as from Practical Paranoia iOS 12 to Practical Paranoia iOS 16), and these require a new purchase. But if at any time you wish to update to the latest Kindle version of *Practical Paranoia iOS 16* at no cost:

1. Delete the copy of *Practical Paranoia* that is installed on your Kindle device.

2. Download the current edition of *Practical Paranoia* from your Kindle library.

1.7.3   Practical Paranoia *Live!* Online Updates

If you have the Practical Paranoia *Live!* Online edition, you always have the latest update for your OS.

1. Launch your web browser.

2. Go to your Google Drive portal (*https://drive.google.com*) > *Shared With Me*.

3. The new version is found in your *Practical Paranoia Textbook* folder.

1.8   Note for Trainers

The *Live!* online edition is a streaming pdf version of the book. We update as the technology changes, often every month or two. We strongly recommend opting for the *Live!* online digital version of the Practical Paranoia books. These allow the students and teacher to always have the most current version—at a lower price than either the paperback or kindle versions.

Please contact our office for details:

Email: *info@thepracticalparanoid.com*
Voice: +1 505.453.0479

1.9   Update Bounty

Although we work tirelessly to keep the contents of this workbook up to date, every now then something slips by. If you discover anything in this workbook that does not reflect current reality, and you are the first to report it to us, we thank you, and you will receive a free signed copy of *Practical Paranoia.*

To make an update bounty report:

Email: *info@thepracticalparanoid.com*

1.10 [Windows] Running Programs on Windows

1.11 Format Conventions Used in this Book

Italics are used to represent hyperlinks, action steps, and file and folder locations.

`Courier` is used to represent user input in a command-line environment.

- **Warning** is displayed when an action is potentially destructive to your data.

- Prerequisite is displayed when additional resources are required to complete an assignment.

Footnotes with hyperlinks are displayed to provide either additional, deeper, or original source information.

NA in a heading indicates a chapter, section, or assignment that is used in one of the other books in the series, but not for this particular book or OS.

1 Titles are chapter headings.

1.1 Sections are section headings.

1.1.1 Assignments are assignment headings.

1.12 Threats and Vulnerabilities

Your cybersecurity and internet privacy is threatened from dozens of points of vulnerabilities–far more than even the professionals are aware. This is because in this cat-and-mouse game, there are an unknown number of unknown points of attack. These are called zero-day vulnerabilities.

Putting aside zero-day issues for the moment, let's examine the more popular vulnerabilities each of us face daily:

Your Computer, Tablet, and Phone

- *Weak passwords*. Should a bad actor get their hands on your device with a weak password, they now have the keys to your kingdom and access to your credit cards, bank accounts, everyday communications, and dark secrets.

- *Passwords easily bypassed.* Should your device lack strong encryption, even if it has a strong password, the password can simply be bypassed.

- *Reusing Passwords.* Because it is so difficult to remember strong passwords, most people use the same few passwords for every site and service. But if one of these accounts is hacked, the criminals now have your credentials which may work on dozens of sites.

- *Malware.* Malware can gain a foothold to your device from email, website, file sharing, text message, and application. There have even been a few incidents where a legitimate antivirus application downloaded from the developer was the carrier of malware.

- *Fingerprinting.* Your device, with its combination of OS version, device model, installed fonts, memory, etc. is unique, giving it a unique fingerprint. This fingerprint is used by websites and trackers to follow your internet activities. Following your activities generates a precise profile of your likes, dislikes, political, psychological, financial, even sexual preferences.

Wi-Fi, Ethernet, and Cellular Connection Router

The first step your device makes out to the local area network or internet is through the local network transport.

- *Ethernet.* While sending data through an ethernet cable, the cable acts as a broadcast antenna. Off-the-shelf receiving devices can pick up the broadcasts, reading all your traffic from miles away.

- *Wi-Fi.* Most Wi-Fi networks are either unencrypted or are using encryption protocols that have been cracked. This leaves all your communications not much more private than sending smoke signals.

- *Router Password.* Most routers still use the factory default administrator username and password. This allows a bad actor administrative access to your entire network.

- *Out of Date Router Firmware.* Just as with the software on your device, the firmware on a router must be updated to patch security holes. But very few are ever updated.

- *Cellular.* Although all cellular signals are encrypted, the encryption protocol was cracked long ago.

Internet Routers

Your communications leave your router then pass-through internet routers on the way to your Internet Service Provider (ISP). First stop is the Domain Name Service (DNS) hosted at your ISP.

- *Internet Routers.* Your internet activity can be monitored and recorded at any of the dozens of internet routers that pass along your communications.

- *DNS.* This service records each server you visit, adding to your personal profile. It is common practice for your profile to be sold to marketing groups.

Web Server

- *Monitoring and Fingerprinting.* Most websites monitor all visitors to an extreme degree. Combine this with fingerprinting and passing profile information from site to site, these organizations now literally know more about you than your mother.

- *Email.* Although you may know that your email service is encrypted, do you know if your recipient can say the same? Email containing sensitive personal or proprietary business information can easily be harvested.

- *Hacked Accounts.* Even if you use a strong password for a website, what if the site does not securely store your credentials and are then hacked. It is now common for hundreds of millions of accounts to be compromised at once.

Voice, Video, and Instant Messaging

- *Insecure Communications Software.* Voice, video, and instant messaging dominate our digital communications. But almost all the applications used store, monitor, and monetize your communications.

Social Media

- Please, do not get me started on this. It's just too early in the course.

1.13 What is at Stake

In my workshops there are always a few participants who proclaim, "I have nothing to hide, I do not really care about cybersecurity and internet privacy".

My universal reply is, "would you please write the following on the whiteboard:

- Your full legal name
- Your residential address
- Your phone number
- Your Social Security Number
- Your credit scores
- Your income
- Your computer and phone passwords
- Your passwords for Facebook, LinkedIn, Google, Microsoft, and Apple
- Your log in credentials for your bank and credit cards"

Nobody yet has taken me up on the challenge. Not because we have something to hide (although some of you do – you know who you are), but because if this information fell into the wrong hands, it would be damaging to you.

Cybersecurity and internet privacy are critically important because your finances, your relationships, your career, your company, and perhaps your life are dependent upon keeping some information out of the hands of bad actors.

The good news is it is easy to do once you know how, and we are here to guide you step-by-step.

1.14 iOS Version

The world of cybersecurity and internet privacy is a 24/7/365 cat-and-mouse game. Criminals, including state agencies, have massive time and labor resources to spend discovering vulnerabilities to systems.

As we will discover later in this course, one of the primary reasons for system and app updates and upgrades is to patch security and privacy vulnerabilities. Simply put, if your system and apps are not current, there are far more opportunities for penetration than if all are current.

1.14.1   Assignment: Verify iOS Version

In this assignment, you verify your iOS version. Later in the course you will learn how to set automatic system and application updates.

1. Tap *Settings > General > Software Update.*

2. If there is an iOS update available, follow the onscreen instructions to download and install. When installation completes, continue.

3. Tap *Settings > General > About.* Your device *Software Version* displays.

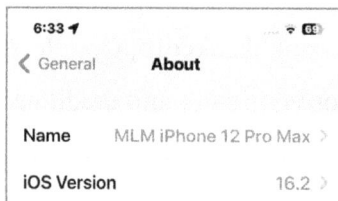

6:33	⚡
‹ General	**About**
Name	MLM iPhone 12 Pro Max ›
iOS Version	16.2 ›

4. If the *Version* is at least *16,* you are good to go. If the *Version* is lower than *16,* your device is not designed for the current OS. It is time to replace the device.

1.15   Legal

Cybersecurity and internet privacy is a non-stop cat-and-mouse game. If you have your devices and systems configured to the best of industry standards today, it is entirely possible that you will have a vulnerability – and perhaps a breach – tomorrow.

In addition to all the sources of vulnerabilities listed within this course, we also face:

• Software updates changing a security setting or creating a zero-day vulnerability.

- System updates changing a security setting or creating a zero-day vulnerability.

- New malware entering your device that is not recognized by your anti-malware software.

- A bug in software or system.

- Human error (you or someone using your device).

- Someone (friend, family, guest) accessing your network via their own compromised device.

The Practical Paranoid (TPP) and the authors have endeavored to provide the most effective solutions based on industry best practices to cybersecurity and internet privacy for the typical individual, household, and small to medium sized business. However, neither TPP, its agents, or the authors are your full-time cybersecurity and internet privacy consultants. By keeping this book or participating in this course, you agree to hold harmless TPP, its agents, and authors from any damages, real or otherwise.

1.16 What Is New in iOS 16 Cybersecurity and Internet Privacy

Cybersecurity and Internet Privacy issues[10] that are new or significantly changed from iOS 15 to iOS 16 include:

- DNSSEC:
  - o Domain Name Server Security Extensions (DNSSEC) provides built-in DNS security, ensuring apps only talk to who they are designed to be talking to.

- Face ID:
  - o Face ID now works in landscape mode.

- Family Sharing:

---

[10] *https://www.apple.com/ios/ios-16-preview/features/*

- o Create accounts for children with parental controls.
- Mail:
  - o You can unsend a recent email.
- Messages:
  - o You can now edit a message you just sent.
  - o You can unsend a recent message.
- Passkeys (FIDO2 passwordless initiative):
  - o Passkeys are intended to replace passwords.
  - o Passkeys are a form of public-private key encryption, with a private key held securely on the user's device and a public key on the server or website.
- Paste Permissions:
  - o Apps now need user permission before access the pasteboard to copy/paste from another app.
- Photos App Security
  - o Sensitive or deleted photos are secured from unauthorized viewing. Two new Photos app albums–*Hidden* and *Recently Deleted*–unlock only with Face ID, Touch ID, or a passcode.
- Private Access Tokens:
  - o Intended to eliminate the need for CAPTCHAS (once implemented within websites).
- System Security and Privacy:
  - o Lockdown Mode add extreme security for users who may be subject to a targeted attack.
  - o Manage Sharing & Access Setting displays an overview of what you are sharing so that you can stop unwanted tracking or monitoring.
  - o Pasteboard permission requires your permission before an app can access the pasteboard to paste content from another app.

- o Rapid Security Response applies system security updates faster, usually without need for restart.

- o Safety Check helps those in domestic violence situations review and reset the access they've granted others. Can reset system privacy permissions for apps and restricts Messages and FaceTime to the device on hand.

- Wallet:

  - o Wallet can share your keys securely with messaging apps, including Messages, Mail, and WhatsApp.

  - o You control when and where your keys can be used, and you can revoke them anytime.

  - o Wallet can be used to present your ID within apps to verify your identity.

1.17 Introduction Lessons Learned

- ☐ Cybersecurity and Internet Privacy are *everyone's* business, not just for medical, banks, and ISPs.

  - ☐ Maintaining cybersecurity and privacy are important to everybody who has a personal life and anything of value.

  - ☐ There are very serious reasons to ensure security and privacy because you never know who may track everything you do and give access to others.

  - ☐ CISPA allows the US government easy access to all your electronic communications.

  - ☐ PRISM allows the US government to collect and track data on any American device.

  - ☐ Zero-day exploits are vulnerabilities not yet discovered by the developer, so your malware detector may not save you.

  - ☐ It does not require extensive training and expertise to protect yourself. Everything you need to know is in this guide.

☐ Keeping system and apps up to date is an important step to ensuring cybersecurity and internet privacy.

☐ New to iOS 16 is Lockdown Mode, which provides extreme security for users who may be subject to targeted attack.

2 Data Loss

Weather forecast for tonight: Dark.

–George Carlin[12]

I know, you want to jump right into cyber security and harden your awesome device. Sorry to be a Debbie Downer[13], but there is a very real risk of losing data in the process of some of the work ahead of us. Because of this, we must begin our exciting journey into the heart of security with drudgery: backing up your computer.

What You Will Learn in This Chapter

- Appreciate your need for backups
- iOS backup to computer
- Verify iOS computer backup
- iOS data recovery from computer
- iOS backup to iCloud
- Verify iOS backup to iCloud
- iOS data recovery from iCloud

What You Will Need in This Chapter

- [Optional] A computer with free space adequate to hold a backup of your device.

---

[12] *https://en.wikipedia.org/wiki/George_Carlin*
[13] *https://en.wikipedia.org/wiki/Debbie_Downer*

- [Optional] iCloud+ with storage space adequate for backing up your device. $1 or higher.

2.1   The Need for Backups

Data loss is a very real fact of life. It is not a matter of *if* you will experience data loss, just a matter of when, and how often. Only a small percentage of computer users back up on a regular basis. I suspect these are the folks who have experienced catastrophic data loss and never want a repeat.

There are many sources of data loss. The top contenders include:

- Device theft
- Power surges
- Power sags
- Sabotage
- Fire
- Water damage. I personally have had 3 clients who have lost computers due to cats or dogs marking their territory, and my own cat took out a $4,000 monitor with nothing more than a hairball.
- Entropy / aging of the hardware
- Malware
- Terrorist activities
- Criminal activities
- Static electricity
- Physical shock to the device (banging the device, dropping, etc.)

Best practice[14] calls for at least two backups:

- **One full backup onsite**. This allows for almost immediate recovery of lost or corrupted documents, or full recovery of the OS, applications, and documents in the event of catastrophic loss.

- **One full backup offsite**. This is your *Plan B* in the event of a catastrophic loss of both the device and the onsite backup. This typically takes the form of fire or theft.

- **One Internet-based backup**. This is your OMG, what do I do now? Fallback plan. Many people substitute the Internet backup for the offsite. A potential problem is that your Internet backup may take several days to weeks to download.

Onsite Full Backup

Fortunately, Apple has built-in the ability to perform both local and Internet-based backups. You may opt for either backup solution, preferably both.

The local backup is performed through the Finder (macOS) or iTunes (Windows). The advantage is that depending on the speed of your computer and iOS device, it may complete in a matter of minutes. The disadvantage is a high vulnerability to loss of the backup through fire, theft, etc. To protect your computer data from such loss, US-CERT and I strongly recommend daily internet-based backups.

Offsite Full Backup

An offsite full backup contains the same content as the onsite backup, but after the backup completes, the backup storage device is store offsite. Offsite backups should be performed at least once per week, preferably daily.

An external storage device, typically a flash drive or SSD is used to back up the computer used to back up your iOS device. Both macOS and Windows included built-in software to perform a full backup. The full off-site backup requires knowledge of how to perform a full backup of your computer. This is covered in detail in *Practical Paranoia macOS Security Essentials,* and *Practical Paranoia Windows Security Essentials.*

---

[14] *https://en.wikipedia.org/wiki/Best_practice*

Once the computer backup completes, the external storage device is removed and transported to a secure storage location. This is typically a bank safe deposit box or a trusted neighbor.

Internet-Based Backup

The Internet backup is performed with iCloud. Once iCloud backup is enabled, the device will automatically attempt a backup to iCloud when the device is turned on, locked, and connected to a power source. The advantage is minimal vulnerability to loss. The disadvantage being that many people aren't thrilled with handing their data over to any corporate entity for safekeeping.

What's Included in an iOS Backup[15]

Item to Backup	Computer Backup	iCloud Backup
Data that is already stored in iCloud, i.e.: Contacts, Calendars, Mail, Music, Notes, My Photo Stream, iCloud Photo Library	No	No
Data stored on other cloud services, i.e.: Gmail and Exchange mail	Yes	No
Content synced from iTunes, like imported MP3s, CDs, videos, books, and photos	No	No
Activity, Health, Keychain data	No-if not encrypted Yes-if encrypted	No
App Data	Yes	Yes
Apple Pay information and settings	No	No
Apple Watch backup	Yes	Yes
Face ID or Touch ID Settings	No	No
Call history	No	No
Camera Roll	Yes	Yes
Device settings	Yes	Yes
Encryption	Off by default May be enabled	Always
iMessage, text, and MMS messages	Yes	Yes

---

[15] *https://support.apple.com/en-us/HT204136*

Item to Backup	Computer Backup	iCloud Backup
Photos and videos on your iPhone, iPad, and iPod touch	Yes	Yes
Synched Photo Libraries	No	No
Media not purchased on iTunes	Yes	No
Purchase history from Apple services (music, movies, TV shows, apps, and books)	Yes	Yes
Ringtones	Yes	Yes
Storage	Limited by computer	5GB default Up to 2TB for fee
Visual Voicemail password (requires the SIM card that was in used during backup)	Yes	Yes
Voice Memos	Yes	No

2.1.1   [macOS and Windows] Assignment: Format the Backup Drive

2.1.2   Assignment: iOS Backup To Computer

Backup to macOS

In this assignment, you configure your iOS device to back up to your macOS 11 or higher computer. If you are using a Windows machine, skip to later in this assignment.

How often should you back up? This depends on how often your data changes, if you also back up to iCloud, and your pain threshold to data loss. As a broad generalization, daily wouldn't be too often.

1. Connect your iOS device to your macOS computer using an iOS data sync cable.

2. If prompted on your iOS device if you trust this computer, tap *Yes*. If prompted on your computer if you trust this iOS device, tap *Yes*.

3. Select the name of your iOS device in the sidebar of a Finder window, in the *Locations* area.

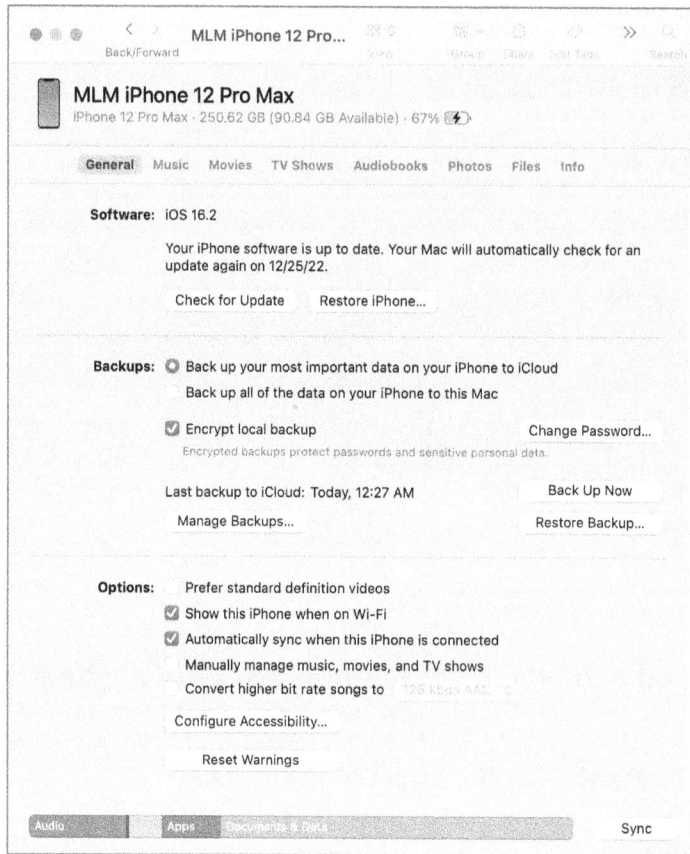

4. In the Finder window > *Backups* area, select *Back up all the data on your iPhone to this Mac*.

5. In the Finder window > *Backups* area, enable *Encrypt local backup*.

6. Tap *Back Up Now* button.

7. If this is the first time this device is backed up to this computer, at the prompt, enter a password to encrypt the backup, then securely store the password.

8. When the backup completes, tap the *Done* button.

9. If you leave your iOS device connected to your computer, it automatically backs up daily.

How often should you back up? This depends on how often your data changes, if you also back up to iCloud, and your pain threshold to data loss. As a broad generalization, daily wouldn't be too often.

Backup To Windows

In this assignment, you configure your iOS device to back up to your Windows 11 computer.

1. If you do not have the iTunes app installed, visit the Microsoft Store at *https://www.microsoft.com/p/itunes/9pb2mz1zmb1s?rtc=1&activetab=pivot:o verviewtab.*

2. Connect your iOS device to your Windows computer using an iOS data sync cable.

3. Launch iTunes.

4. Tap the *Device* button near the top left of the iTunes window.

5. Tap Summary.

6. Below the *Backups* area, enable *Encrypt backup.* At the prompt, enter a password to encrypt the backup then securely record it.

7. Tap *Backup Up Now.*

8. If you leave your iOS device connected to your computer with iTunes open, it automatically backs up daily.

2.1.3   Assignment: Integrity Test Backup to a Computer

Murphy lives in technology. If anything can go wrong, it tends to do so. No different for backups.

In this assignment, you verify your iOS backup to your computer.

1. Connect your iOS device to your computer.

2. On a Mac, open a Finder window, then select your iOS device from the sidebar. On a Windows computer, launch iTunes, select the *Device* icon near the top left of the window, then select the target iOS device.

3. In the *Backups* area, note the date of Last backup to this computer <date>.

4. If the date matches when you last expected a backup, all is good.

2.1.4 [Optional] Assignment: Restore iOS Device from a Computer Backup[16]

Oh, *Snap!* Your iPhone or iPad has up and walked away, jumped into the deep end of the pool, or served well as a chew toy for the new puppy. The good news is that a bright, shiny, new iOS device is waiting to make your acquaintance. The better news is that your data is safely backed up and ready to snuggle up with your that new device.

In this assignment, you restore your data to either a new iOS device, or your existing one that has been wiped clean.

- **Warning**: Only perform this assignment if you need to restore your iOS device.

- Prerequisite: Completion of a current, full back up to your Mac or Windows computer.

1. Attach the iOS device to the computer holding its backup.

2. On a Mac, open a Finder window, then select the iOS device from the sidebar. On a Windows computer, launch iTunes, select the *Device* icon near the top left of the window, then select the target iOS device name.

3. Tap the *Restore iPhone...* button.

4. At the Are you sure you want to restore the iPhone alert, tap the Restore button.

5. Follow the few onscreen instructions to complete the restoration.

Restoring a device may take an hour or longer to complete, but when done, the new phone is a duplicate of the other at the time of backup.

2.1.5 [Android] Assignment: Smart Switch Backup to Local Storage

---

[16] *https://support.apple.com/en-us/HT204184*

2.1.6 [Android] Assignment: Smart Switch Restore from Local Storage

2.1.7 [Windows] Assignment: Encrypt File History Backup

2.1.8 Assignment: Internet Backup with iCloud[17]

Every Apple ID comes with 5 GB of free storage on iCloud. Additional storage space can be purchased. iOS ships with free built-in backup software. When using iCloud backup, you can have most of your iOS data synchronized with iCloud. This makes iCloud backup a good solution for some individuals.

Due to the lack of HIPAA, SEC, and NIST compliance, we don't recommend iCloud for business backup.

In this assignment, you configure your iOS device to back up to iCloud when connected to power, locked, and connected to a Wi-Fi network with Internet access.

1. On your iOS device, select *Settings > Your Name > iCloud > iCloud Backup*, then slide the iCloud Backup switch to On.

---

[17] *https://support.apple.com/en-us/HT204184*

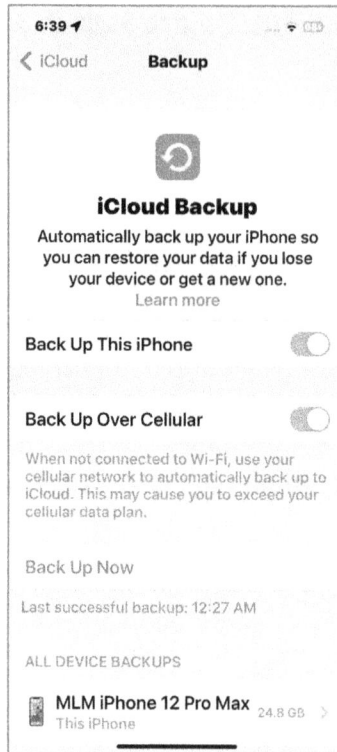

2. To force a backup now, select *Back Up Now.* The *Backup* screen indicates a backup is in progress, and how long until complete.

You may continue to work while your device backs up.

2.1.9   Assignment: Integrity Test iCloud Backup

Unfortunately, there is no integrity test that can be performed to validate an iCloud backup. But there is a quick and easy way to verify it is backing up your files.

In this assignment, you verify your iCloud backup is backing up your files.

- Prerequisite: Completion of an iCloud backup.

1. On your iOS device, go to *Settings > your name > iCloud > iCloud Backup.*

2. At the bottom of the screen is a status report. In this example: *Last Successful Backup: 9:17 am.* I'm current. If the status was two or more days past or reported something like: *The last backup could not be completed*, you have a problem to investigate.

3. While you are here, select *Back Up Now* so that your backup is up-to-the-minute current.

4. Close *Settings*.

2.1.10   [Optional] Assignment: Restore Files From iCloud

You may also use an iCloud backup of an iOS device to restore to another iOS device. Unlike a computer restore, no apps are restored with an iCloud restore. Instead, you need to manually download these from the App Store.

In this assignment, you restore your data from an iCloud backup.

● Prerequisite: Completion of an iCloud backup.

● **Warning**: This assignment will erase the current contents of your device. Unless you need to recover your data, skip this assignment.

1. On a new iOS device, or one that has been erased through *Settings > General > Transfer or Reset iPhone > Reset*, upon startup the *Setup* screen appears. Tap *Restore from iCloud Backup.*

2. At the *iCloud Sign In* screen, enter your *Apple ID* and *Password,* then tap *Next.*

3. You are prompted to authenticate with the *Verify Identity* screen. Tap your desired verification method.

4. When you have received your verification code, enter it in the *Enter Code* screen.

5. At the *Terms and Conditions* screen, gather your legal team to descipher the contract, then tap *Agree.*

6. You are greeted with the *Hello* screen. Slide the screen to set up your backup.

7. At the *Choose backup* screen, tap the desired backup to use in restoring this iOS device. Unless it is damaged, you want the most recent backup.

8. Restoration begins. Depending on the model of your iOS device, speed of Internet connection, and volume of data to be restored, this process may take from a couple minutes to an hour.

9. When restoration is complete, tap *App Store* app to download your apps.

2.1.11   [Android] Assignment: Backup to a Computer with Dr. Fone

2.1.12   [Android] Assignment: Integrity Test Backup to a Computer with Dr. Fone

2.1.13   [Android] Assignment: Restore Device from Dr. Fone

2.1.14   [Windows] Assignment: Encrypt File History Backup.

2.1.15   Assignment: Internet Backup with Google Drive

Google offers internet-based backup that is encrypted in transit (uploaded from your device to Google and downloaded from Google to your device), encrypted at rest (on the Google server), point in time recoverable, cross-platform, includes file sharing, and with a business-class account will sign a Business Associate Agreement to be HIPAA-compliant.

At this time, we do not recommend using Google Drive to back up your iPhone or iPad, as iCloud provides a superior solution.

Full details may be found at: *https://support.google.com/drive/answer/7070690?hl=en&co=GENIE.Platform%3DiOS*

- The free version is limited to 15GB. Additional storage may be purchased, upgrading to the Google One license.

- The Google One[18] version starts at $1.99-$9.99/month for 100GB-2TB storage, the option to add up to 5 family members, with access to Google support.

- Business-Class[19] license starts at $4.20-$18.00/month for 30GB-5TB storage, with access to Google support and additional security options.

- iPhone data will be backed up to Google Drive, photos will back up to Google Photos, Contacts will back up to Google Contacts, and Calendar will back up to Google Calendar.

- Your iOS device must be on Wi-Fi to back up photos and videos.

- If your iOS device has photos organized into albums, those albums will not back up to Google Photos.

- Contacts and calendars from services like Facebook, Exchange, Apple, will not back up.

2.1.16 Assignment: Integrity Test Google Backup

In this assignment, you verify the integrity of your Google Backup.

1. Open Google *Drive.app.*

2. Tap the *Notifications* tab.

3. Verify there are no messages regarding synchronization errors.

4. Close *Drive.*

2.1.17 Assignment: Data Recovery from Google

Unlike with Android, there is no "data recovery" when backing up iOS to Google. Should you acquire a new device, simply installing the Google Drive app, and then authenticating in it, will give you access to all your data.

---

[18] *https://one.google.com/about*

[19] *https://workspace.google.com/intl/en_id/business/*

2.2   [Windows] Recovery Drive

2.3   Data Loss Lessons Learned

☐ Data loss is a fact of life with many sources of loss including theft, power surges and sags, sabotage, fire, water damage, entropy, malware, physical shock, static electricity, criminal, and terrorist activities.

☐ Best practice calls for at least two backups, at least one onsite and at least one offsite.

☐ As the owner of an iPhone or iPad, you have a free Apple iCloud account, or can create an iCloud account.

☐ Data that is already stored in iCloud is not backed up with either a computer backup or iCloud backup.

☐ Backups should be integrity tested monthly.

☐ Google Backup and Sync is a free backup service and application, limited to 15 GB data storage for all free Google accounts, and either 1 TB or unlimited data storage for all paid Google accounts.

3 Passwords

For a people who are free, and who mean to remain so, a well-organized and armed militia is their best security.

–Thomas Jefferson[1]

Knowledge, and the willingness to act upon it, is our greatest defense.

–Marc L. Mintz[2]

What You Will Learn in This Chapter

- Create a strong password
- Use the Keychain
- View an existing Keychain record
- Create challenge questions
- Store challenge Q&A
- Access secure data from Keychain
- Synchronize Keychain across macOS and iOS devices
- Use Bitwarden to save website credentials
- Create password policies
- Perform 2-factor authentication

---

[1] *https://en.wikipedia.org/wiki/Thomas_Jefferson*
[2] *https://thepracticalparanoid.com*

What You Will Need in This Chapter

- [Optional] $10 for Bitwarden yearly subscription.

3.1   The Great Awakening

In June 2013, documents of NSA origin were leaked to The Guardian newspaper[3]. The documents provided evidence that the NSA was both legally and illegally spying on United States citizens' cell phone, email, and web usage. These documents, while causing gasps of outrage and shock by the public, revealed little that those of us in the IT field already did not know/suspect for decades: every aspect of our digital lives is subject to eavesdropping.

The more cynical amongst us go even further, stating that *everything* we do on our computers *is* recorded and subject to government scrutiny.

But few of us have anything real to fear from our government. Where the real problems with digital data theft come from are local kids hijacking networks, professional cyber-criminals who have fully automated the process of scanning networks for valuable information, competitors/enemies and malware that finds its way into our systems from criminals, foreign governments, and our own government.

The first step to securing our data is to secure our computers and mobile devices. Remember, we are not in Kansas anymore.

3.2   Strong Passwords

We all know we need passwords. But do you know that *every* password can be broken? Start by trying *a.* If that does not work, try *b.* Then *c.* Eventually, the correct string of characters gets you into the system. It is only a matter of time. Most attackers have more time than you do.

---

[3] *https://en.wikipedia.org/wiki/NSA_warrantless_surveillance_controversy*

Way back in your great-great-great grandfather's day, the only way to break into a personal computer was by manually attempting to guess the password. Given that manual attempts could proceed at one attempt per second, an 8-character password became the standard. With a typical character set of 26 (a–z) this created a possibility of 26^8 or over 100 billion combinations. The thought that anyone could ever break such a password was ridiculous, so your ancestors became complacent.

A while back the first hacker wrote password-breaking software. Assuming it may have taken eight CPU cycles to process a single attack event, on an old computer with a blazing 16 KHz CPU that equated to 2,000 attempts per second. This meant that a password could be broken in less than 2 years.

All of this is funny when you consider that research has shown most passwords can be guessed. Most computer users are unaware that what they thought was an obscure and impossible-to-break password could be cracked in minutes.

Yikes. IT directors took notice.

Down came the edict from the IT Director that we *must* create *obscure* passwords: strings that include upper and lower case, numeric, and symbol characters. But in many cases, this was a step backward. Since a computer user could not remember their password was 8@dC%Z#2, the user often would manually record the password. That urban legend of leaving a password on a sticky note under the keyboard? I have seen it myself more than a hundred times.

In 2003, Bill Burr came up with a set of password security guidelines for NIST. These included requirements to include a number, a capital letter, and a symbol like an asterisk. Unfortunately, that guidance was wrong. If you have an 8-character password, but one letter must be a capital, then that one letter has only 26 possible values instead of 72 (26 lower-case letters, 26 upper-case, 0-9 and 10 special characters). You haven't made your password safer – you've reduced the *key space* and made it easier to crack! Require a number, and that character only has 10 possible values – making your password easier to crack still. Require a special character (assuming the 10 most common such as @) and your password is yet easier to crack. In 2017, Mr. Burr issued a massive apology[4] because his advice had made passwords weaker, not stronger.

---

[4] *https://www.wsj.com/articles/the-man-who-wrote-those-password-rules-has-a-new-tip-n3v-r-m1-d-1502124118*

Present day: A current personal computer with freely available password-cracking software can make over 10 billion password attempts per second. Create an army of infected computers called a botnet to do your dirty work[5] and you can achieve over a hundred trillion attempts per second, unless your system locks out the user after x number of failed logon attempts. That's critical, but while local applications and network logons are likely to use a preset lockout rate (say, three unsuccessful attempts), many web sites and online applications do not.

What does this army of enemy computers mean for you? The typical password using upper and lower case, number, and symbol now can be cracked with the right tools in under two minutes. If the hacker uses just a single computer to do the break in, make that a whole week. Don't believe it? Look at the *Bitwarden Password Strength Testing Tool*[6].

It's true that if we use longer passwords, we can make it too time consuming to break into our system, so bad guys may move on to other targets. But you say it is tough enough to remember *eight* characters, impossible to remember more?

That's only true if we keep doing things as done before. Since attacks now are done by automated software, password hacking is only an issue of length of password, not complexity. So, forget memorizing meaningless combinations of digits and use a passphrase that is easy to remember, such as, "Rocky has brown eyes" which at 100 trillion attempts per second could take over 1,000,000,000,000,000 centuries to break (provided Rocky is not the name of your beloved pet and thus more guessable).

How long should you make your password, or rather, passphrase? My recommendation to clients is a minimum of 15, in an easy-to-enter phrase. As of this writing, Apple[7] and Microsoft[8] recommend a minimum of 8 characters, while Google[9] recommends 12. US-CERT[10, 11] recommends at least 15 for

---

[5] *https://en.wikipedia.org/wiki/Botnet*

[6] *https://bitwarden.com/password-strength/*

[7] *https://support.apple.com/en-us/HT201303*

[8] *https://www.microsoft.com/en-us/research/wp-content/uploads/2016/06/Microsoft_Password_Guidance-1.pdf*

[9] *https://support.google.com/a/answer/33386?hl=en*

[10] *https://security.web.cern.ch/security/recommendations/en/passwords.shtml*

[11] *https://www.us-cert.gov/ncas/alerts/TA11-200A*

administrative accounts, at least 8 for non-administrators. Cisco recommends[12] at least 8.

In addition to password length, it is critical to use a variety of passwords. In this way, should a bad actor gain access to your Facebook password, that password cannot be used to access your bank account. There's the famous story of Mat Honan[13], who had worked for Gizmodo and WIRED Magazine (he should have known better, but this was back in 2012 when dinosaurs still roamed), whose iCloud account was hacked. That's bad enough, but Mat had used the same password on Gmail, Twitter, and other accounts – all of which also were hacked in less than an hour.

Yes, soon you have a drawer full of passwords for all your different accounts, email, social networks, financial institutions, etc. How to keep all passwords organized and easily accessed amongst all your various computers and devices? More on that later in this chapter.

Microsoft and Apple are encouraging users to move away from passwords and move toward fingerprint, facial recognition, and other (possibly more secure) options. It would be prudent to mention that biometrics are only more secure against remote attacks. In person, someone could get control of your finger, they could use it to access the machine. Same goes for your face. May seem unlikely but there are cases before the courts of law enforcement officers doing just that and more to get biometric access to systems without a warrant. As a bonus, the courts have been inconsistent[14] when ruling on the legality and constitutionality of said cases.

Our current recommendation is to NOT use biometrics to log in or wake any device, but once logged in, it is acceptable to use biometrics for purchases, and as an alternative to entering passwords.

---

[12] *https://www.cisco.com/c/en/us/td/docs/ios-xml/ios/sec_usr_aaa/configuration/15-sy/sec-usr-aaa-15-sy-book/sec-aaa-comm-criteria-pwd.html*

[13] *https://www.wired.com/2012/08/apple-amazon-mat-honan-hacking/*

[14] *https://9to5mac.com/2020/08/12/police-demand-you-unlock/*

2023 Apple Password Recommendations[15]

- Minimum 8-character length

- Include at least one number

- Include both upper and lowercase letters

- For a stronger password, add additional characters

2023 Google Password Recommendations[16]

- Use a different password for every site

- Minimum 12 characters long

2023 Microsoft Password Recommendations[17]

- Minimum 8-character length

- Contain characters from at least three of the following: Uppercase letters, lowercase letters, base 10 digits, special characters, any Unicode character that is categorized as an alphabetic character but is not uppercase or lowercase

- Must not contain the user's account name value

- Must not contain the user's display name

- Eliminate character-composition requirements

- Eliminate mandatory periodic password resets for user accounts

- Ban common passwords, to keep the most vulnerable passwords out of your system

- Educate users not to re-use a password for non-work-related purposes

- Use multi-factor (2-factor) authentication

---

[15] *https://support.apple.com/en-us/HT201303*

[16] *https://support.google.com/accounts/answer/32040?hl=en#zippy=%2Cmake-your-password-unique%2Cmake-your-password-longer-more-memorable*

[17] *https://docs.microsoft.com/en-us/windows/security/threat-protection/security-policy-settings/password-must-meet-complexity-requirements*

2023 US-CERT Password Recommendations[18, 19]

- Private and known only by one person

- Not stored in clear text in any file or program, or on paper

- Easily remembered

- Minimum 15-character length for administrators, minimum 8-character length for non-administrators

- A mixture of at least 3 of the following: upper case, lower case, digits, and symbols

- Not listed in a dictionary of any major language

- Not guessable by any program in a reasonable time frame

A 6-digit passcode (the default for iOS devices) is easy to break. Start by entering *000000,* then *000001,* then *000002,* etc., then after a maximum of 1,000,000 attempts, the bad guy is in.

To help slow down the hacker, your device will temporarily disable after multiple wrong password attempts:

- 6 wrong attempts: Disabled for 1 minute.

- 7 wrong attempts: Disabled for 5 minutes.

- 8 wrong attempts: Disabled for 15 minutes.

- 9 wrong attempts: Disabled for 60 minutes.

- 10 wrong attempts. Disabled until connected to Finder (macOS) or iTunes (Windows), then follow onscreen unlocking method.

There are two options to improve the security of your login:

- Create a more complex (longer or alphanumeric) passcode.

- Configure the device to automatically erase after 10 failed passcode attempts.

---

[18] *https://www.us-cert.gov/ncas/tips/ST04-002*
[19] *https://www.us-cert.gov/ncas/alerts/TA11-200A*

- Note: Every child over 3 whose parent has an iPhone or iPad has discovered this.

3.2.1   Assignment: Create a Strong Password

In this assignment, you test the strength of a password for your device or website.

Test password strength

1. Think up a password for yourself that consists of at least 15 easy-to-enter characters and meets the strength/complexity required by your organization.

2. Open a web browser to *https://bitwarden.com/password-strength/*. Scroll down to the *Evaluate Your Password* entry field. Then enter your password in the field.

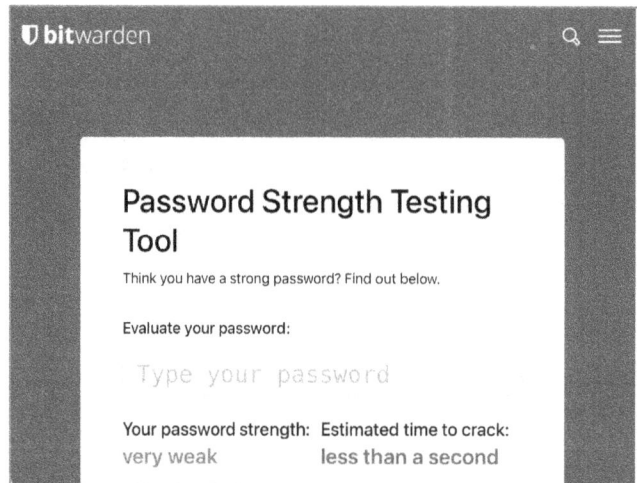

3. Scroll down to the *Your password strength* and *Estimated time to crack*.

4. Repeat with several more of your passwords.

5. If you find passwords that do not meet your security needs, change them now.

6. Record your new password.

7. Exit the browser.

Change your old password to the strong password

8. If the password is for a website, visit it now and change the password.

3.2.2 [Optional] Assignment: Create a New Login Password

In this assignment, you configure your iOS device to use the default *Simple Passcode.* This is a 6-character password consisting only of numbers. Combined with the following assignment, *Enable Erase Data on Failed Passcode Attempts,* your iOS device should meet any password security compliance.

1. Select *Settings > Face ID* or *Touch ID & Passcode.*

2. At the *Enter your old passcode* screen, enter your current passcode.

3. Scroll down then tap *Change passcode.*

4. Enter your current passcode.

5. In the *Enter your new passcode* screen, tap *Passcode Options.* Your options are:

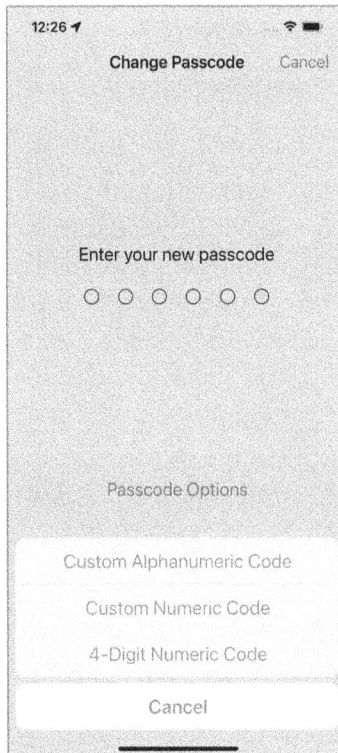

6. If you wish to have a longer or more complex passcode, select the first or second option, then enter your desired passcode. For this assignment you create a new 6-digit numeric code, so tap *Cancel*.

7. At the *Enter your new passcode* screen, enter your desired 6-digit passcode.

8. At the prompt, re-enter your new passcode to verify.

9. Exit out of Settings.

Congratulations! You have just done more to secure your device than most users!

3.2.3   Assignment: Erase Data After 10 Failed Passcode Attempts

In this assignment, configure the device to automatically erase after 10 failed passcode attempts.

1. Tap *Settings > Face ID* or *Touch ID & Passcode*.

2. At the prompt, enter your passcode.

3. Scroll to the bottom of the screen to enable *Erase Data*.

- This makes two changes to the operation of your device. First, if someone attempts to hack your passcode, the device fully auto erases after the tenth attempt. Second, they likely never get to the tenth attempt, as after five attempts, the device locks for 1 minute. After the next failed attempt, the device locks for 5 minutes, then 15 minutes, and so on.

4. Exit *Settings*.

3.2.4   [Windows] Assignment: Sign-In Options

3.3   Password Manager - Keychain

In our grandparent's day, life was so much simpler. To give an example: My grandfather always had four keys in his pocket: one for home, one for the car, and the other two he could never remember what for.

In today's world, the realm of keys has expanded into the digital world. You now have keys or passwords for logging onto your computer, phone, tablet, email, many of the websites you visit, Wi-Fi access points, servers, frequent flyer account, etc. In my case, I have 857 passwords in use. I know because they are all neatly stored in a database so that I do not have to remember them.

Unfortunately for most of us, our "keys" are not very well organized, so when we need to access our mail from another computer, or order a book on Amazon, we are stuck.

By default, your iOS device stores most usernames and passwords used to access Wi-Fi networks, servers, other computers, and websites. The exceptions are usually websites that are programmed specifically so they do not have credentials saved. These are typically financial institutions.

The place where your iOS device stores this information is called the *Keychain*. The Keychain database is secured with industry-standard AES 256 encryption.

When visiting a website that requires a username and password, connecting to another computer or server, or performing some other action that triggers an authentication request, the following steps typically occur:

1. A prompt appears requesting a username and password.

2. You enter your username and password.

3. The website takes you to the appropriate page.

Behind the curtain, your device has copied your username and password into the Keychain database.

The next time you visit this same website, the steps change somewhat:

1. You surf to the website.

2. A prompt appears requesting a username and password.

3. Behind the scenes your web browser asks: Has the Keychain stored the credentials for this site?

4. A query is made of the Keychain database.

5. If Keychain has stored the username and password associated with the URL or server (it has), the credentials are automatically inserted into the username and password fields.

6. Select *Enter*.

7. The website takes you to the appropriate page.

Note that you did not need to know your credentials–Keychain did it all for you.

● New as of iOS 15 and macOS 12 is the ability to share Keychain passwords with Windows users with iCloud for Windows, and iCloud Passwords browser extension for Chrome and Edge.

3.3.1 [Android] Assignment: Enable Autofill with Google

3.3.2 Assignment: View an Existing Keychain Record

For this assignment, let us assume you have forgotten the password to your Wi-Fi network. The Keychain database has stored the password. You just need to look for it.

In this assignment, you examine a record in the Keychain.

1. On your iOS device, tap *Settings > Passwords*. A list of all saved passwords displays.

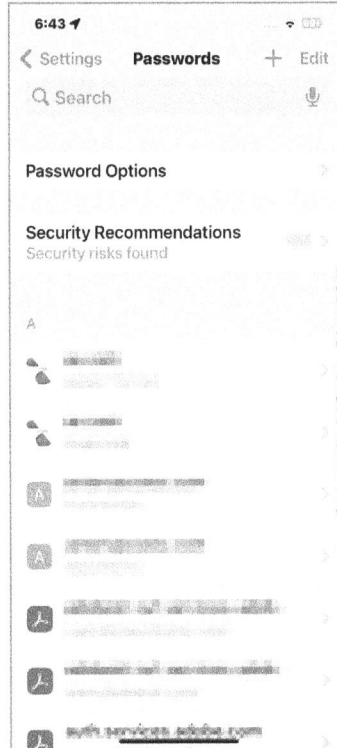

2. Tap on the name of the target website or server. The credential records display.

3. Tap on *Password* to display the password in cleartext.

4. Exit *Settings*.

3.3.3   Assignment: Keychain Auto-Entry

In this assignment, you have Keychain auto-enter username and password for a website.

- Prerequisite: Website credentials stored in Keychain.

1. Open Safari on your iOS device, then go to a website that requires credentials (username and password). For this example, I am going to *https://appleid.apple.com.*

2. Tap in the *username* field. Keychain displays all stored options just above the keyboard. In my example, I have multiple accounts for this site. The first two display.

3. In this example, the account I wish to use isn't displayed, so I tap the *Key* icon to list all accounts in Keychain for this site.

4. Tap the desired account.

5. The username is auto filled.

6. Tap the continue icon (->).

7. Tap in the *Password* field. Keychain displays the available accounts for this site. The first one matches the username entered earlier in the assignment.

8. Tap the account to log in with. The password field is auto filled with the password.

9. Tap the *Continue* icon (->) to log in.

3.3.4   [Optional] Assignment: Harden Keychain Security

The Keychain remains unlocked while the device is unlocked. This means that if you leave your device behind in class, the airport screening area, or restroom, it is very likely it be unlocked when someone finds it. And that someone then has access to all your credentials.

The solution is to set your *Auto-Lock* (the amount of time the device can go unused before a passcode is required to unlock) to as short a time as comfortable.

In this assignment, you configure your Auto-Lock.

Face ID & Passcode

This setting determines how long after the screen goes dark does the screen lock.

1. Open *Settings > Face ID & Passcode* (or *Touch ID & Passcode)*.

2. Enter your device passcode.

3. Scroll down to *Require Passcode* area, then tap on the timer pop-up menu.

4. Select the shortest time comfortable for you.

Auto-Lock

This setting determines how long before an unattended screen goes to black.

5. Open *Settings > Display and Brightness > Auto-Lock*.

6. Configure to your taste. It recommended 1 minute or less.

7. Close *Settings*

3.3.5 Assignment: Use Passwords System Setting to View Passwords

In this assignment, you explore the *Passwords* System Setting.

View Existing Password

1. Open *Settings > Passwords*. The Keychain opens, displaying accounts with stored credentials.

2. Tap on a target account. The account record opens.

3. Tap in the *Password* field. The password for this account changes from bullets to cleartext.

Change a Password

4. Tap the *Edit* button at top right.

5. Tap on the target password.

6. You may now change the password.

3.3.6 Assignment: View Credit Card Information

The Keychain also stores credit cards for use in Safari, Apple Pay and Wallet. In this assignment, we view any credit card information stored on your iOS device.

1. Select *Settings > Wallet & Apple Pay*.

2. At the authentication prompt, enter your device password or Face/Touch ID.

3. Any credit cards that have been saved appear.

4. Select one of the cards. Full card information displays.

5. Exit from *Settings*.

Congratulations! You now know more about the iOS keychain than you ever wanted to know.

3.4   Challenge Questions

A Challenge Question is a way for websites to authenticate who you claim to be when you contact support because of a lost or compromised password.

For example, when registering at a website you may see an entry for: *Question – Where did your mother and father meet?*

The problem with this strategy is that most answers are easily discovered with an Internet search of your personal information, or a bit of social engineering.

The solution is to give bogus answers. For example, my answer to the question; *Where did your mother and father meet?* might be: *1954 Plymouth back seat.* It would not be possible for a hacker to discover this answer, as it is completely bogus. My mother tells me it was a 1952 Dodge.

Unless you are a savant, there is no way to remember the answers to all challenge questions. But there is no need to remember. We already have a built-in utility that is highly secure and designed to hold secrets such as passwords: Keychain Access!

Although Keychain can automatically record and auto fill usernames and passwords, and on macOS it allows for manual entry of data such as challenge Q&A, there isn't a good way to accomplish this in iOS Keychain. For that, we need to use a third-party password manager utility, such as *Bitwarden.* More on that in a few pages.

3.4.1   [macOS] Assignment: Store Challenge Q&A in the Keychain

3.5   Synchronize Keychain Across iOS, macOS, and Windows Devices

Perhaps like me, you may need to access most of these passwords anywhere, anytime. It also is a huge help to have the credentials created on your macOS computer available to your iOS device, and vice versa.

If you have macOS X 10.9 or higher or macOS and iOS 7 or higher, to a large degree Apple has you handled. With the most recent incarnations of both operating systems, Apple has added *Keychain* to the iCloud synchronization scheme. This allows your Keychain database to be synchronized between all your Apple computers, iPhones, and iPads. New with iOS 15 and macOS 12 is Keychain synchronization with *iCloud for Windows* and Google Chrome and Microsoft Edge browser extensions.

3.5.1   Assignment: Activate iCloud Keychain Synchronization

In this assignment, you enable iCloud Keychain synchronization between all your iOS and macOS devices.

1. Tap *Settings > Your Name > iCloud > Passwords and Keychain*. If the *Sync this iPhone* is *On*, you are already set. If it is *Off*, tap enable.

2. Exit *Settings*.

Your iOS device Keychain now synchronizes with your iCloud account. Enabling iCloud Keychain on your other iOS and macOS devices allows all of them to synchronize the Keychain.

3.6   Password Management with Bitwarden

A great solution to the need for credential, challenge question, and Time-based One-Time Password (TOTP) management is the open-source utility Bitwarden[20]. Bitwarden features the following:

- Secure password sharing
- Security audit and compliance

---

[20] *https://bitwarden.com*

- 24/7 support
- Cross-platform operation
- Cloud-based or self-hosted options
- Fully functional free version, as well as for-fee version
- Two-Factor Authentication one-time password generator (requires for-fee version)

3.6.1   Assignment: Install and Configure Bitwarden

In this assignment you install the free version of Bitwarden.

- Note: If you use only Apple devices and browser, this assignment is optional, as the major advantage of Bitwarden is the ability to work cross platform and with browsers other than Safari.

Download and Install

1. Open *App Store,* search for then download and install Bitwarden.
2. Open *Bitwarden*, then tap *Create Account.*
3. Enter the requested information to create an account, then tap *Submit.* This returns you to the login screen.

Enable AutoFill

4. Open then log in to Bitwarden.
5. Select *Settings > Password AutoFill.*
6. Open *Settings.app.*
7. Tap *Touch ID* or *Face ID and Passwords.*
8. Tap *AutoFill Passwords.*
9. Enable *AutoFill Passwords.*
10. Enable *Bitwarden*
11. Exit *Settings.*

Enable Bitwarden Browser Extension

12. Open *Bitwarden > Settings > App Extension*.

13. Tap *Enable App Extension > Bitwarden*. The *Extension Activated* screen opens. Tap *Back* then *Close* to return to Settings.

Enable Automatic Sync

14. In *Bitwarden Settings* tap *Sync*.

15. Turn on *Enable sync on refresh*, then tap *Sync Vault Now*. When done, tap *Close*.

Set Vault Timeout

16. In *Bitwarden Settings* set *Vault Timeout* to 15 minutes. When done, return to *Settings*.

Set Unlock with Touch ID or Face ID

17. In *Bitwarden Settings* set *Unlock with Touch ID* or *Unlock with Face ID* to Enabled. Then return to *Settings*.

Enable Two-Step Login

As the keys to your treasure are stored in this database, not only is a strong Bitwarden password important, but so is having Two-Step Login (2-Factor Authentication) enabled.

18. Open a browser to *https://bitwarden.com*.

19. Tap *Log In,* enter your login credentials then tap *OK*.

20. From the Bitwarden toolbar, tap *Settings > Two-Step Login*.

21. Select your preferred method to get a verification code. In this assignment you use *Email*.

22. At the prompt, enter your Bitwarden email address, then tap *Send Email*.

23. Open your email to find the verification email.

24. Copy the verification code from your email, paste it into the verification field, then tap *Enable*.

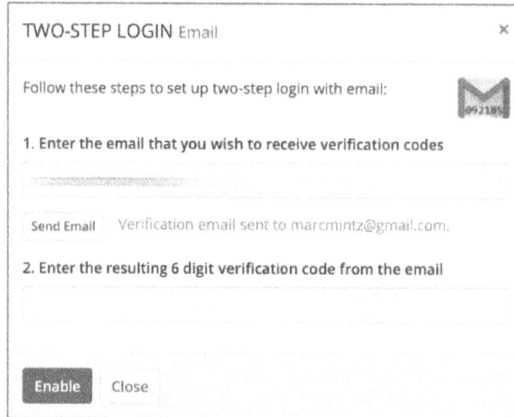

25. At the dialog box stating that This two-step login provider is enabled on your account, tap Close.

26. In the Bitwarden *Two-step Login* webpage, tap *View Recovery Code.*

27. Enter your Bitwarden password, then tap *Continue.*

28. In the *TWO-STEP LOGIN Recovery code* window, copy then securely store your recovery code. When done, tap *Close.*

- Note: In the event you no longer have access to your Bitwarden Authenticator, this is the only way you can regain control over your account.

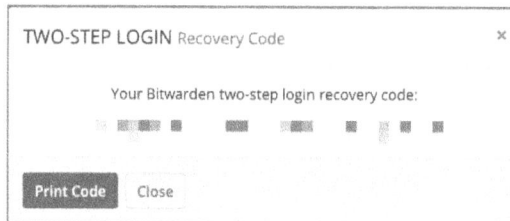

3.6.2   Assignment: Enter New Credentials in Bitwarden

Credentials may be entered either manually or automatically into Bitwarden.

In this assignment you manually enter new credentials into Bitwarden.

- Prerequisite: Bitwarden account, Bitwarden installed on your device.

Manual Entry

1. Open *Bitwarden > Vault*.

2. Tap + in the top-right corner.

3. In the *Type* field, select the type of entry (*Login, Card, Identity, Secure Note*).

4. Enter login credentials for *Name, Username, Password.*

5. Enter the website (for example, *thepracticalparanoid.com*) in the *URI* field.

6. Optionally, set the *Folder* to store this record, record any notes, turn on *Favorite*.

7. When complete, tap *Save.*

Automatic Entry

8. Visit a website that is not stored in Bitwarden.

9. Enter your site login credentials, then tap *OK.*

10. Bitwarden pops up a dialog asking if you want to save these credentials. Tap *Yes.*

3.7   Password Policies

A password policy is a set of rules to help users create and use passwords. You have seen password policies in use when creating a password for your online banking or shopping and were alerted that your password needed to be longer or have a special character.

Within the government, military, financial, and healthcare environments, and the NIST 800-171 security protocols, setting *password policies* is a mandate. Although not a mandate for the home and general business computer, doing so makes a lot of sense.

In an IT environment which is controlled by a Microsoft Active Directory server, password policies can be enforced from the server. In environments without a server, password policies can be enforced using a Mobile Device Management service.

Active Directory server and Mobile Device Management are topics outside the scope of this course and will not be covered here.

3.7.1 [macOS] Assignment: Password Policies with Apple Configurator

3.7.2 [macOS] Assignment: Install Password Policies Profile

3.8 2-Factor Authentication

Maintaining strong passwords that are unique to every site is the foundation for site and account security. But it is only the start.

Every password can be cracked. Many websites do not properly encrypt passwords. When such a site is itself hacked, millions of user account and password credentials are harvested in seconds.

The fix for these issues is the use of *2-Factor Authentication (2FA)*, sometimes referred to as *Multi-Factor Authentication* or *2-Step Authentication.* With 2FA active, not only are the account username and password required for access, but a second form is also required. This can take the form of:

- An email.
- SMS or text message of a code to your phone.
  - o Note: This is no longer considered high security, as it is possible to intercept such text messages.
- 2FA random code generation performed by an app.
- 2FA random code generation performed by a hardware device (typically a USB stick).

There are dozens of 2FA apps available, with the *Google Authenticator* being the most popular. However, my recommendations go to *Bitwarden*[21] and *Passwords Settings* (iOS 15 and higher). Their benefits include:

- Easy 2FA setup for any account

- High security, with Touch ID, Face ID, and PIN protection

- Cloud-based encrypted backup

- Use on multiple devices

- Bitwarden works cross platform with Android, Chrome OS, iOS, macOS, and Windows.

- Password works with macOS 12 and higher and iOS 15 and iPadOS 15 and higher.

- Note: Passwords only works with Safari.

- Note: Bitwarden is open-source and free, but to enable its Authenticator module (as well as a few other features) requires upgrading to a for-fee tier, ranging from $10/year to $40/year.

3.8.1 [Optional] Assignment: Configure Bitwarden Two-Step Authentication

In this assignment, you enable Bitwarden 2-Factor Authentication for a website. This is normally referred to as 2FA, 2-Step Authentication, 2-Step Verification, Multi-Factor Authentication, and in the case of Bitwarden *Authenticator Key (TOTP)* for *Time-based One-Time Password)*. For this assignment we use Bitwarden to setup two-step authentication for a Google account.

- Prerequisite: Installation of Bitwarden, and purchase of one of its premium tiers.

1. Open a browser and go to *https://security.google.com.*

2. If prompted, authenticate.

3. In the main body area, scroll down to tap on *2-Step Verification.*

---

[21] *https://bitwarden.com*

4. At the prompt, enter your Google credentials to authenticate again, then tap *Next.*

5. Scroll down to the *Authenticator app* section, then tap *SET UP.*

6. At the *Get codes from the Authenticator app,* select the type of smartphone you use (Android or iPhone), then tap *Next.*

7. In the *Set-up Authenticator* window, it is designed to be captured with a smartphone camera.

If Using a Browser on a Mobile Device

a. Tap *CAN'T SCAN IT?*

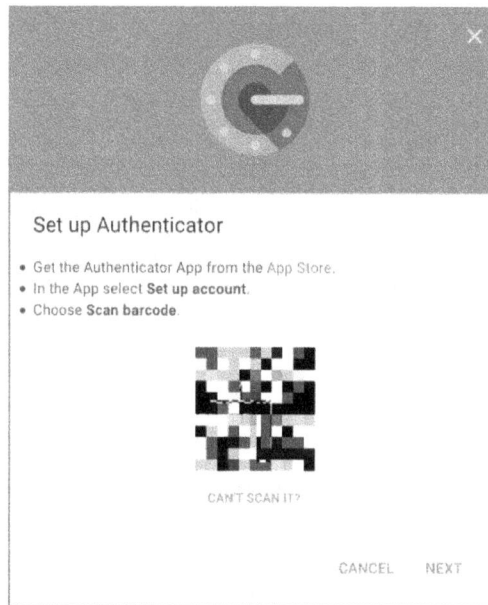

b. In the *Can't scan the barcode?* dialog, note the 32-character code.

c. On your mobile device, open *Bitwarden > Vault,* select your Google account, then tap the *Edit* (pencil) icon.

d. Enter the code from step 7b into the *Authenticator Key (TOTP)* field, then tap the *Save* (disk) icon.

If Using a Browser on a Computer

a. On your mobile device, open *Bitwarden > Vault,* select your Google account, then tap the *Edit* (pencil) icon.

b. In the Authenticator Key (TOTP) field, tap the Camera icon.

c. Point your camera at the QR code displayed on the computer screen. Bitwarden automatically recognizes, captures the code, then enters the numeric value in the field.

d. Tap Save.

8. In the *ITEM INFORMATION* area of your Bitwarden Google record, you now see a *Verification Code (TOTP)* field. This is the one-time only authenticator code that can be used when prompted by Google. If you have other devices with Bitwarden, they now also have this new field.

3.8.2 Assignment: Configure 2FA in Passwords Settings

In this assignment, you configure 2FA using the built-in *Passwords Setting*.

1. Open Safari and go to *https://security.google.com.*

2. If prompted, authenticate.

3. In the main body area, scroll down to tap on *2-Step Verification.*

4. At the prompt, enter your Google credentials to authenticate again, then tap *Next.*

5. Scroll down to the *Authenticator app* section, then tap *SET UP*.

6. At the *Get codes from the Authenticator app,* select the type of smartphone you use (Android or iPhone/iPad), then tap *Next.*

7. In the *Set-up Authenticator* window, it is designed to be captured with a smartphone camera. Assuming you are using your smartphone to view the page…

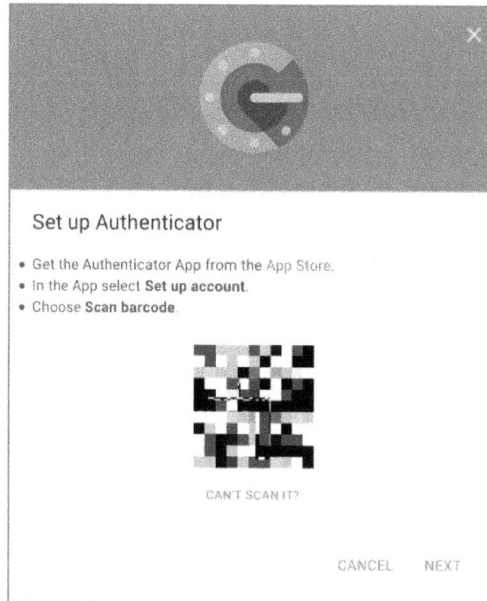

8. Tap *CAN'T SCAN IT?*

9. In the *Can't scan the barcode?* Dialog, record the 32-character code.

10. On your device, open *Settings > Passwords.*

11. Open your Google record.

12. Tap *Edit* button > *Enter Setup Key… .*

13. Enter the 32-character code displayed at step 9, then tap *OK.*

14. The random 2FA Verification Code will display within the Google record in Passwords.

15. The next time you are requested for the 2FA code to access Google, if it is not auto entered, tap in the code field and Passwords will autofill the 2FA code.

3.9   Passwords Lessons Learned

☐ Documents of NSA origin, acquired by Edward Snowden and leaked to The Guardian newspaper in 2013, provided evidence the NSA was legally and illegally spying on US citizens' cell phone, email, and web usage.

☐ Each website and service should have its own unique password.

☐ The first measure of a passwords strength is a minimum of 15 characters.

☐ The second measure of a passwords' strength is a mix of upper and lower case characters, numerals, and special characters.

☐ A unique password should be created for every site and service.

☐ Keychain is the iOS built-in password manager.

☐ Keychain database can be shared and synchronized across all Apple devices sharing the same Apple ID account.

☐ Challenge Questions should not have Answers that are true.

☐ Bitwarden is a cross-platform password manager and Time-based One-Time Password (TOTP) Authenticator.

☐ Bitwarden requires a web browser extension to communicate with browsers.

☐ A Password Policy is a set of rules to help users create and use passwords.

☐ Two-Factor Authentication (2FA) resolves the vulnerability of cracked passwords.

☐ 2FA requires a secondary verification in the form of an email, SMS or text message, code generated by an application, or code generated by a hardware device.

☐ New as of iOS 15 and macOS 12 is the ability to share Keychain passwords with Windows users with iCloud for Windows, and iCloud Passwords browser extension for Chrome and Edge.

☐ New as of iOS 15 is the Passwords Setting ability to create, store, and autofill 2FA codes.

4 System and App Updates and Privacy

Every new beginning comes from some other beginning's end.

–Seneca[1], Roman philosopher, statesman, and dramatist

What You Will Learn in This Chapter

- Configure iOS system and application update schedule
- Provide system and application privacy

What You Will Need in This Chapter

- No additional resources required

4.1   System and App Updates

Most computer and mobile device users simply fail to update their systems. Many say the reason is that updates slow down the device, or they are concerned about introducing instability to their systems.

Updates rarely significantly change performance, although upgrades often do because they have a larger code base, requiring more RAM and CPU. While it is occasionally true that updates introduce instability, it is far more likely that updating creates greater stability.

More important is that many updates and upgrades patch vulnerabilities and security holes in a system. Fixing these security issues is so important that US-CERT (Homeland Security division responsible for cyber terrorism and IT

---

[1] *https://en.wikipedia.org/wiki/Seneca_the_Elder*

security) strongly recommends that all users update all computers and mobile devices "as soon as possible."[2, 3, 4]

There are fundamentally three reasons for updates and upgrades:

- **Bug fixes**. All software and hardware have bugs. We never will be rid of them. Developers do want to squash as many as possible so that you are so happy with their product and continue to pay for upgrades.

- **Monetization**. Updates to operating systems and applications are usually free. Upgrades are for fee. But developers may include significant new features in an upgrade to encourage the market to purchase. Purchases are necessary so developers can afford to stay in business.

- **Security patches**. Although rarely talked about, one of the most important reasons for an update is to patch newly discovered security holes. Without the update, your computer may be highly vulnerable to attack today even if yesterday everything was secure.

It is for this last reason alone that I implore clients to be consistent with the update process. To protect your computer from security holes, it is critical to check for operating system and application updates daily. Fortunately, we can automate this process.

By default, the iOS and apps update nightly when the device is locked, connected to the internet, and connected to power.

4.1.1   Assignment: Configure System and App Updates

iOS has the option to automatically update music, apps, books & audiobooks, and iOS. You may also manually update.

In this assignment, you update your iOS device system software and apps.

- Prerequisite: Your device must be plugged into power, or the battery must be fully charged.

---

[2] *https://www.us-cert.gov/ncas/tips/ST04-006*

[3] *https://nvlpubs.nist.gov/nistpubs/SpecialPublications/NIST.SP.800-167.pdf*

[4] *https://www.cisecurity.org/critical-controls/documents/TheASD35andCISControls.pdf*

- Recommended: Back up prior to any system update, due to the small, but real, potential for catastrophic corruption during this process. See chapter: *Data Loss.*

Configure automatic iOS and app updates

1. Tap *Settings > General > Software Update.*

2. Tap *Automatic Updates.* The *Automatic Updates* screen appears.

3. In the *Automatic Updates* screen, enable *Download iOS Updates, Install iOS Updates, and Security Responses & System Files,* then tap *Back* and exit *Settings.*

4.1.2 Assignment: Manually Update iOS and Apps

It is not uncommon for the iOS and apps to go without updating for days or weeks, even with updates available. As this presents a potential security and stability vulnerability, I recommend manually checking for updates at least monthly.

In this assignment, you manually check for updates.

Check for and Update System Software

1. Open *Settings > General > Software Update*.

2. If the iOS is up to date, exit *Settings*. If the *Software Update* screen shows there is an update available, tap the *Update* button, then follow the on-screen instructions to complete installation.

3. Exit *Settings*.

Check for and Update Apps

4. Open *App Store*.app.

5. Tap your avatar in the top right of the screen.

6. Swipe down (starting an inch or two from the top) to force a refresh.

7. Scroll down to *Upcoming Automatic Updates*.

8. If there are updates available, tap *Update All*.

9. App updates begin downloading and installing.

10. Exit *App Store.app*

Congratulations! Your iPhone or iPad is now updated and secure from more malware, bugs, and penetration attempts than a few minutes before. But do not rest on that information. It appears evil never sleeps, and it is already working away on new ways to compromise your data.

4.1.3   [macOS] Assignment: Install and Configure MacUpdater

4.1.4   [Android] Assignment: Update all Apps

4.1.5   [Android] Assignment: Require Authentication for App Purchases

4.2   System and App Privacy

System and App developers often need feedback on how their users use their product. However, we as the users, must first be respectful of our own privacy, and give a thumbs-up or down to letting the system, application, and developer look over our shoulders.

4.2.1   Assignment: Configure System and App Privacy

In this assignment you configure the Privacy settings of your iOS or iPadOS device. There are no correct set of settings, and each person needs to determine the proper balance of privacy/flexibility/functionality that works for them.

Enable Location Services

1. On your device, go to *Settings > Privacy & Security > Location Services.* The *Location Services* screen opens.

2. Enable *Location Services.* Location Services allows applications to know their geographical location.

3. In the *Location Services* screen, configure each application for how you wish to share your location.

4. Exit *Settings.*

Share My Location

5. Open the *Find My* app.

6. Tap *People* icon *> Start Sharing Location* button.

7. In the *Share My Location* screen, select who you wish to share your location, then tap *Send.*

8. Select how long you want to share your location.

9. When your friend or family member opens their Find My app, they see your location.

10. Tap *<Back* to return to the main *Privacy & Security* screen.

Enable apps to request to track

11. In the *Settings > Privacy & Security* main window tap *Tracking.*

12. Tap the *Allow Apps to Request to Track.* Apps then request the ability to track your location across other apps and websites. This is primarily used by advertisers to link all your web activity. But it can also be used by others to know all about you. For instance, this tool was used to know who participated in the January 6, 2021, Capitol riot, then track their exact locations at all times starting from when they left their home states. A rogue app may still be able to track you even if denied or the switch is disabled. When done, tap *Privacy* to go back.

13. Tap *<Back* to return to the main *Privacy & Security* screen.

Configure Analytics & Improvements

14. In the *Privacy & Security* main screen, scroll down to select *Analytics & Improvements.*

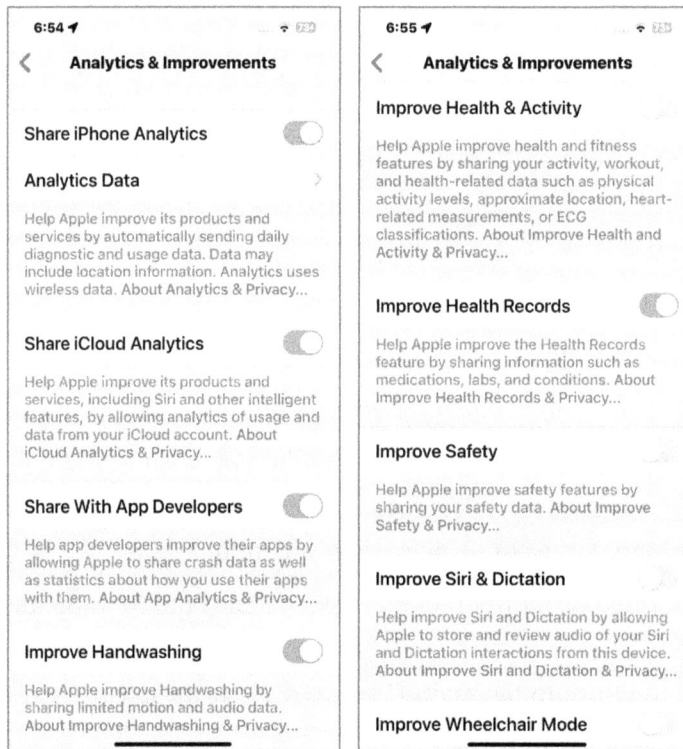

15. If you want to share how your iPhone is used with Apple, tap the *Share iPhone Analytics* switch.

16. If you want to share your app use with the developer, tap the *Share With App Developers* switch.

17. If you want to improve Siri and Dictation, tap the *Improve Siri & Dictation* switch.

- Note: This sends all your Siri and Dictation audio files to Apple for analysis.

18. If you want to share you use iCloud with Apple, tap the *Share iCloud Analytics* switch.

19. If you want to share your workout and health-related data with Apple to improve their *Health Records* apps, tap the *Improve Health Records* switch.

20. If you use an Apple Watch and want to send your handwashing data to Apple to improve their health-related analytics, tap the *Improve Handwashing* switch.

21. If you use a wheelchair and want to send your wheelchair activity and workout data to Apple to help improve health and fitness features, tap the *Improve Wheelchair Mode* switch, then tap *Back* to return to the main *Privacy & Security* screen.

Apple Advertising

When using Safari, you can allow or prevent Apple from collecting your browsing and other activity for Apple to provide personalized advertising.

22. In the main *Privacy & Security* screen, scroll down to select *Apple Advertising.*

23. If you want Apple to collect your browsing and other activity to help personalize advertising, enable the *Personalized Ads* switch, then tap *Back* to return to the main *Privacy & Security* screen.

Record App Activity

Introduced with iOS 15 is the App Activity Report. This option presents a summary of the times your apps collected your data over a seven-day span. This report includes:

- Contact
- Domains contacted by the app
- Location

- Microphone

- Photos

24. Open *Settings > Privacy & Security > App Privacy Report.*

25. Enable *Turn On App Privacy Report.*

26. Tap *<Back* to return to the main *Security & Privacy* screen.

27. After at least one day of recording, you can return to this page to view app activity.

28. You can download or send yourself the recording by tapping the *Save App Activity* button on the *Settings > Privacy > Record App Activity* page. The file is in .json file format. This can be opened in TextEdit or any text editing application. However, it is not intended for human viewing. To help make more sense of the record, visit the App Store to install *App Activity Viewer.*

29. Exit *Settings.*

4.2.2   [macOS] Assignment: Configure Gatekeeper

4.2.3   [Android] Assignment: Clear Your Local Search History in Play Store

4.2.4   [Android] Assignment: Secure Play Store from Harmful Apps

4.3   System and App Updates and Privacy Lessons Learned

☐ System and app updates are an essential part of security and privacy.

☐ System and app updates exist to monetize the product, implement bug fixes, and eliminate potential and existing security vulnerabilities.

☐ iOS and iPadOS update automatically nightly when plugged into power, connected to the internet, and screen locked.

- ☐ iOS and iPadOS automatic updates are configured from Settings > General > Software Update.

- ☐ App Store is configured from Settings > App Store.

- ☐ Privacy settings are configured from Settings > Privacy & Security.

- ☐ New as of iOS 15 is the ability to view the App Privacy Report.

4.4   Additional Reading

Souppaya, Murugiah, and Karen Scarfone. "Guide to Enterprise Patch Management Technologies." NIST Special Publication 800-40, Revision 3. July 2013. <https://nvlpubs.nist.gov/nistpubs/SpecialPublications/NIST.SP.800-40r3.pdf

Liu, Simon, Rick Kuhn, and Hart Rossman. "Surviving Insecure IT: Effective Patch Management." Insecure IT. 2009. <http://csrc.nist.gov/staff/Kuhn/liu-kuhn-rossman-v11-n2.pdf

5 User Accounts

Those who desire to give up freedom in order to gain security will not have, nor do they deserve, either one.

–Benjamin Franklin[1]

What You Will Learn in This Chapter

- Understand user accounts
- Understand application Whitelisting
- Configure Screen Time

What You Will Need in This Chapter

- No additional resources required.

5.1   User Accounts

User accounts allow you to compartmentalize your data so that it is not exposed to those who may be using your device. You may end up doing all the other security steps to harden your device to the core, but otherwise your data could be exposed when your friend uses your device for 5 minutes to "check my email." If you have the same users commonly using your device, then it is a very good idea to assign them their own user accounts. By assigning a user a separate user account, that user cannot access your specific data such as pictures, app-specific data, etc. Each

---

[1] *https://en.wikipedia.org/wiki/Benjamin_Franklin*

new user account has its own authentication method (or none if the user chooses), separate from your own method.

The user accounts feature is of great use for parents or co-owners of a device such as spouses. Separating your workspaces allows for customization and convenience in how the device operates for each user. Also, having your data only within your workspace and locked behind your authentication is critical to protect against accidental data compromise.

- As of iOS 16, Apple has not yet implemented user accounts. Unlike Android, Chrome OS, macOS and Windows, iOS 16 remains a single-user system.

- Included with iOS is *Screen Time.* Screen Time allows activation of some of the features of separate user accounts. It is of particular importance and use when a device is used by children or employees. We cover Screen Time in another chapter.

5.2 [Chrome OS, macOS, and Windows] Never Log In As an Administrator

5.2.1 [macOS] Assignment: Enable the Root User

5.2.2 [macOS] Assignment: Log In as Root

5.2.3 [macOS] Assignment: Change Root Password

5.2.4 [macOS] Assignment: Disable Root

5.2.5 [macOS] Assignment: Create an Administrative User Account

5.2.6 [macOS] Assignment: Change from Administrator to Standard User

5.2.7 [Android] Assignment: Configure or Remove a User Account

5.3 Screen Time Control and Application Whitelisting

In 2014, Target, Home Depot, and other major retailers were hacked for their customer databases. Although there were multiple breakdowns in the security protocols of these organizations, one step would have prevented all of them: *application whitelisting.* This same strategy should be used by both home and business systems to help secure computer systems.

Application whitelisting is a process that allows only authorized applications to run on a computer, blocking any executable not on the list. This is a vital ingredient to system security because even the best anti-malware catches only 99.9% of the *known* bugs. And if your computer is penetrated by *unknown* malware, anti-malware is of no use. However, if your computer has application whitelisting in place, the unknown malware is blocked from executing.

While macOS devices do have the ability for full application whitelisting, iOS devices do not. Due to the built-in security protocols, and vetting of apps on the App store, there has not been a need for whitelisting in iOS.

However, you may wish to restrict access to apps if handing your iOS to someone else to use. *Screen Time* is built into macOS, iOS, and iPadOS devices to do just this.

Screen Time[2] provides a monitor and control portal for all your devices, and those of your staff and children. You can monitor how much time is spent with specific apps, place limits on the amount of time specific apps are available, and to restrict communication to only known people.

5.3.1 [Android] Assignment: Enable Family Link

---

[2] *https://support.apple.com/en-us/HT208982*

5.3.2   [Optional] Assignment: Configure Screen Time

In this assignment, you enable Screen Time. After which, you can activate the various Screen time options.

Setup Family Sharing

This part is optional. If you would like to manage up to six family members, complete this part. Otherwise, skip to the next section *Enable Screen Time*. More information on Family Sharing can be found at *https://support.apple.com/en-us/HT201088*. More information on Screen Time can be found at *https://support.apple.com/en-us/HT210387*.

1. Open *Settings > <Your Name> >Family Sharing > Continue*.

2. Tap *Invite Others* or tap the *Invite* button for the suggested accounts.

3. Follow the on-screen instructions to add the selected users to your Family Sharing account.

Enable Screen Time

4. Go to *Settings > Screen Time*.

5. Tap *Turn On Screen Time*.

6. Tap *Turn On Screen Time* button.

7. At the *Is This iPhone for Yourself or Your Child* screen, select the appropriate option, then follow the on-screen instructions to complete.

* Note: If prompted to create a Screen Time password, record and securely store the password. If you forget the password, the only way to disable Screen Time is to factory reset your device.

5.3.3   [Android] Assignment: Turn On and Configure Play Store Parental Controls

5.3.4   [Android] Assignment: Enable Pin Windows

5.3.5   [Windows] Assignment: Configure a Child Account

5.3.6 [Windows] Assignment: Allow Access to Websites Through Email

5.4 [macOS and Windows] Policy Banner

5.4.1 [macOS and Windows] Assignment: Create a Policy Banner

5.5 User Accounts Lessons Learned

☐ User accounts allow compartmentalization of your data, so the data is not exposed to others using your device.

☐ iOS and iPadOS do not have multiple user accounts.

☐ Screen time provides a monitor and control portal.

☐ Screen Time is enabled from *Settings > Screen Time.*

6 Device Hardware

I am disturbed by how states abuse laws on Internet access. I am concerned that surveillance programs are becoming too aggressive. I understand that national security and criminal activity may justify some exceptional and narrowly tailored use of surveillance. But that is more reason to safeguard human rights and fundamental freedoms.

–Ban Ki-moon[1], Secretary General of the United Nations

What You Will Learn in This Chapter

- Device encryption
- Configure a SIM card lock

What You Will Need in This Chapter

- [Optional] iPad MiFi-compliant case

6.1 [macOS and Windows] Block Access to Storage Devices

6.1.1 [Windows] Assignment: Block Access to Storage Devices Via Registry

---

[1] *https://en.wikipedia.org/wiki/Ban_Ki-moon*

6.1.2   [Windows] Assignment: Restore Access to Storage Devices Via Registry

6.1.3   [Windows] Assignment: Block Access to Storage Devices Via Device Manager

6.1.4   [Windows] Assignment: Restore Access to Storage Devices Via Device Manager

6.2   [Windows] Assignment: Restore Access to Storage Devices Via Device Manager

6.3   Device Encryption

It is important to have your data protected from anyone who has stolen your device or somehow comes upon it on the subway. If someone were to steal your device and attempt to copy the data from it, any unencrypted data are keys to the kingdom for the thief.

iOS 16 automatically enables hardware encryption on your entire storage device as soon as a passcode is assigned. There is nothing else that you need do to.

6.3.1   [macOS and Windows] Assignment: Enable and Configure Storage Encryption

6.3.2   [macOS and Windows] Assignment: Enable Encryption on Non-Boot Storage

6.3.3   [Intel Mac] Assignment: Boot Into Target Disk Mode

6.3.4 [Apple silicon Mac] Assignment: Boot Into Target Disk Mode

6.3.5 [Intel Mac] Assignment: Boot Into Recovery Mode

6.3.6 [Apple silicon Mac] Assignment: Boot Into Recovery Mode

6.3.7 [Intel Mac] Assignment: Boot Into Single-User Mode

6.3.8 [Apple silicon Mac] Assignment: Boot Into Single-User Mode

6.4   Resistance to Brute Force Attack

Apple claims there is no back door or golden key to iOS encryption. If there is a way to hack into a protected volume, only one group is laying claim to it. *Passware*[2] says their software can break into an iPhone. The author has not tested this claim.

There are two other barriers to brute force attack on iOS: Auto-erase on 10 failed password attempts, and delayed password entry.

Erase Data After 10 Failed Passcode Attempts

With this feature enabled, the iOS device fully erases after 10 failed password attempts. Be forewarned, inquisitive children and nasty adults can use this to erase your device.

This feature is enabled in assignment 3.2.3.

Delayed Password Entry

This feature is automatically enabled, with no option to disable. After several failed password attempts, the iOS device forces a delay of 1 minute before allowing another attempt. After another failed attempt, the device forces a delay

---

[2] *https://www.passware.com/*

of 5 minutes before allowing another attempt, with progressively longer delays with each attempt. This makes it impossible to use brute force to access a device.

6.5 [macOS] Remotely Access and Reboot a FileVault Drive

6.5.1 [macOS] Assignment: Temporarily Disable FileVault

6.5.2 [macOS] [Intel Mac and Windows] Assignment: Test the Firmware Password

6.5.3 [macOS] Assignment: Remove the Firmware Password

6.5.4 [Windows] Assignment: Enable Additional UEFI/BIOS Security Features

6.5.5 [macOS] Assignment: Startup Security for Apple silicon Mac

6.6 [macOS and Windows] Firmware Password

6.6.1 [Android and iOS] Assignment: Set Up a SIM Card Lock

6.6.2 [macOS and Windows] Assignment: Test the Firmware Password

6.6.3 [macOS and Windows] Assignment: Remove the Firmware Password

6.6.4   [macOS] Assignment: Startup Security for Apple silicon Mac

6.7   SIM Card Lock

Mobile devices include a SIM card. This card holds data linking to your carrier account. If your device is ever stolen or compromised, all the encryption in the world does nothing against someone taking the SIM card out of your phone, inserting it into their phone, then stealing your expensive cell phone service.

Password protecting a SIM card is called a *SIM card lock*. Keep in mind that if you do set up a SIM card lock, it prompts you to enter this PIN every time that you restart your device, but not when you wake it up from sleep.

- Note: It has become common for mobile devices to have soldered-in SIM cards. It is not possible (or needed) to lock these newer cards.

6.7.1   Assignment: Set Up A SIM Card Lock

In this assignment, you configure a SIM card lock.

1. Select *Settings > Cellular > SIM PIN*.

2. In the *SIM PIN* screen, enable *SIM PIN*.

3. If you do not currently have a SIM PIM set, enter your carrier's default SIM card PIN, then tap *Done*.

- Note: Be careful, if you do not know the password, do not exceed the attempts as it locks you out of your phone until you call your carrier to reset your PIN.

- **Default SIM PIN for US carriers:**
  - AT&T: 1111
  - Sprint: 1234
  - T-Mobile: 1234 or 0000
  - US Cellular: 1234
  - Verizon: 1111

- **Default SIM PIN for UK carriers:**
  - O2: 0000, 5555, or 1234
  - 3: 0000
  - EE: 1111
  - giffgaff: 5555
  - Orange: 1111
  - T-Mobile: 1210
  - Tesco Mobile: 5555
  - Virgin Mobile: 7890
  - Vodafone: 0000

4. Now that the SIM PIN is active and using the carrier's default PIN code, tap *Change PIN*.

5. Enter your current SIM PIN (if this is the first-time adding PIN, this is your cellular carrier default), then select *Done*.

6. Enter a new SIM PIN, then select *Done*. Confirm the PIN, then select *Done*.

7. You are returned to the *SIM PIN* screen. Exit *Settings*.

Congratulations! You have setup a SIM PIN and protected your carrier data and cell phone service from the bad guys.

6.8   Camera and Microphone Recording Indicator and Hardware Disconnect

Camera and Microphone Indicator

New with iOS 14 and iPadOS 14 is a menu bar indicator whenever an app is accessing the device's microphone or camera. An orange dot indicates microphone access, a green dot indicates camera access.

To determine which app is currently accessing your camera or microphone, open the *Control Center* (swipe down from the top right corner of the screen). The app in question will be listed at the top of the screen.

Microphone Hardware Disconnect

Starting in 2020, all iPad models feature a hardware microphone disconnect[3]. When used with an MFi-compliant case, when the iPad is closed, the microphone is physically disconnected in hardware. This makes it impossible for any software, regardless of privilege level, to access the microphone.

6.8.1 Assignment: See the Recording Indicator in Action

In this assignment, you will see the recording indicator in action.

Test Camera Indicator

1. Wake your iPhone or iPad.

16. Look at the menu bar. You should not see any small colored dots.

17. Open any app that accesses the camera. The *Camera* app is one example.

18. Look again at the menu bar. You will now see a green dot, indicating an app is accessing your camera.

19. Assuming you did not know which app is the culprit, open the *Control Center* by swiping down from the top right corner. You will see the name of the offending app at the top of the screen. When done, close the Control Center:

---

[3] *https://support.apple.com/guide/security/hardware-microphone-disconnect-secbbd20b00b/web*

113

Test Microphone Indicator

20. Open any app that accesses the microphone. Siri will do, just say *Hey Siri*.

21. When Siri responds, you will see a yellow dot in the menu bar, indicating an app is accessing the microphone.

22. Open the *Control Center*.

23. At the top of the Control Center screen, you will see the name of the app accessing the microphone.

24. Close Control Center.

6.8.2   [Optional] Assignment: Test Microphone Hardware Disconnect

In this assignment, you verify the microphone hardware disconnect works.

- Prerequisite: An iPad manufactured in 2020 or newer with an MFi-compliant case.

1. Open the case on your iPad.

2. Open *Voice Memos* app (or any app that can record audio).

3. Start recording.

4. Speak a few words, continue speaking while closing the case, then continue speaking while opening the case.

5. Stop recording.

6. Play the recording. Note how when the case is closed, recording stopped, recording a blank space.

7. Quit the recording app.

6.9   Siri Offline Command Processing

New as of macOS 12 and iOS 15 (for iOS and iPadOS devices with an A12 Bionic processor or later) is Siri offline command processing.

Prior to this, all Siri commands were sent via internet to Apple servers for processing–even for commands destined for local implementation, such as *Open camera*.

Now most commands will be processed on the local device. Only commands requiring external resources–such as weather forecasts–will submit the verbal command to Apple's servers.

This goes a very long way to ensuring your privacy.

6.10   Cybersecurity and Privacy Lessons Learned

☐   iOS 16 automatically enables hardware encryption on the entire storage device as soon as a password is assigned.

☐ iOS 16 has very strong resistance to brute force attacks, including automatically erasing data after 10 failed passcode attempts, and delaying password entry.

☐ Mobile devices (except for some of the latest iPhone 14 and higher models) include a SIM card. This card can be password protected, preventing anyone from using the card on their device to access your mobile services.

☐ Camera and microphone access indicators show as green and yellow dots in the menu bar.

☐ The Control Center displays any app accessing the camera and microphone.

☐ iPads manufactured 2020 and newer have built-in hardware microphone disconnect. This works with MFi-compliant cases when the case is closed.

☐ New as of iOS 15 is that Siri commands are processed on the local device, without sending the audio information to Apple. Only commands that require non-local information (such as weather forecast) send the audio to Apple or third parties.

7 Sleep and Screen Saver

Do not take life too seriously. You will never get out of it alive.

–Elbert Hubbard[1], American writer, publisher, artist, and philosopher

What You Will Learn in This Chapter

- Configure Auto-Lock
- Configure Lock Screen Notifications
- Enable Do Not Disturb

What You Will Need in This Chapter

- No additional resources required.

7.1   Require Password After Sleep

When not actively using a powered-on computer, by default it remains on even if the display goes to sleep. It is a trivial task for someone else to sit in front of the computer and access all your data.

On a mobile device, auto-lock is like a screensaver. After a specified amount of time without use, your device darkens its screen, then require a passcode for access.

To help prevent unauthorized eyes on your iOS device, configure it to lock down after a short period of inactivity (1 minute or less), or upon command.

---

[1] *https://en.wikipedia.org/wiki/Elbert_Hubbard*

7.1.1 Assignment: Configure Screen Saver

The *Auto-Lock* feature configures how long your screen remains on until the system requires a correct password entry.

Lock Screen is a feature like a screen saver, configuring how long your screen remains on until the system blacks it out.

In this assignment, you configure your device to Auto-Lock and Lock Screen.

Configure Require Passcode

1. Open *Settings > Touch ID & Passcode* or *Face ID & Passcode* depending on your device.

2. At the *Enter Passcode* screen, enter your iOS passcode.

3. Scroll to and tap *Require Passcode*.

4. In the *Require Passcode* screen, tap the desired time. *Immediately* or *After 1 minute* should meet any security and compliance requirement.

Configure Auto-Lock

5. Open *Settings > Display & Brightness > Auto-Lock*.

6. In the *Auto-Lock* screen, tap the desired time. *1 minute* should meet any security and compliance requirement.

7. Exit *Settings*.

7.2 Lock Screen Notifications

If you've ever had an embarrassing text message show up on the locked screen of your device while meeting with business clients or friends, you know the importance of restricting lock screen notifications. In addition, for some applications this can pose a serious security risk as these notifications on your lock screen may be viewed by anyone.

7.2.1 Assignment: Restrict Lock Screen Notifications

In this assignment, you configure which notifications may be displayed while your device is in lock screen mode.

1. Select *Settings > Notifications*. The Notifications screen appears.

2. Tap *Show Previews,* select when you want alerts to display – *Always, When Unlocked,* or *Never,* then tap *Back.*

3. Each app that can display notifications appears. For this assignment, tap *Messages*. The *Messages* notifications settings appears.

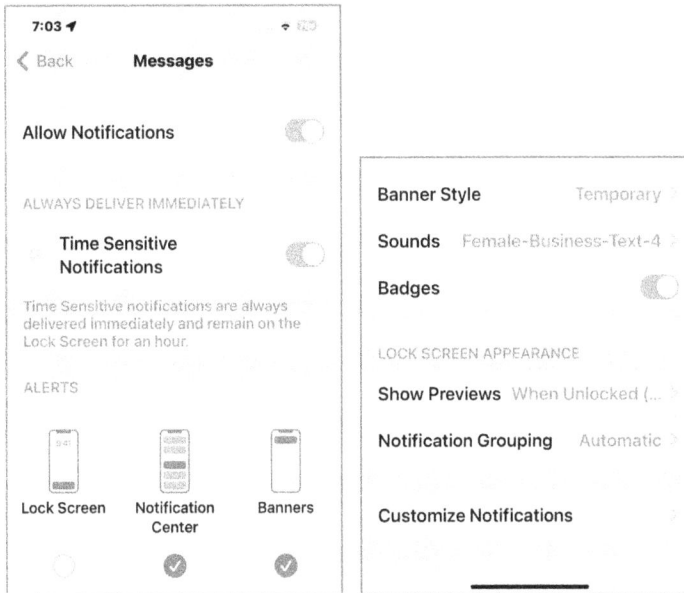

4. To disable an incoming message from appearing on your lock screen, disable *Lock Screen.* Configure all other settings to your taste, then tap *Notifications* to return to the main screen.

5. Configure any other *Notifications* to your taste.

6. Exit *Settings*.

Never again shall you be embarrassed with a message displaying while in lock mode.

7.3   Do Not Disturb Mode

If you ever had your phone go off during an important meeting (or 3am), then you know the value of having *Do Not Disturb mode* in place. This mode allows for you to set times and days when you would like *Do Not Disturb* to be on, as well as what can come through as an exception.

7.3.1   Assignment: Configure Do Not Disturb Mode

In this assignment, you enable *Do Not Disturb*.

1. Select *Settings > Focus*.

2. Tap *Do Not Disturb*.

3. In the *Do Not Disturb* screen, tap *People,* then configure to your taste. Then tap *< Back.*

4. In the *Do Not Disturb* screen, tap *Apps,* then configure to your taste. Then tap *< Back.*

5. In the main *Do Not Disturb* screen, scroll down to tap *Turn On Automatically.*

6. Enable the *Schedule,* then set the *Time* and *Days* to be enabled.

7. Exit *Settings.*

7.4   Sleep and Screen Saver Lessons Learned

☐ It is important to require a password after waking from sleep or screen lock so that passers-by cannot access your device if you are away.

☐ Restrict Lock Screen Notifications from *Settings > Notifications > Show Previews.*

☐ Configure *Do Not Disturb* from Settings > Focus > Do Not Disturb.

8 Malware

Behind every great fortune lies a great crime.

–Honore de Balzac[1], 19th-century novelist and playwright

What You Will Learn in This Chapter

- Anti-Malware
- Outlook.com Safelinks
- Lockdown Mode

What You Will Need in This Chapter

- No additional resources required.

8.1 Anti-Malware

Most people know this category of software as Antivirus. But there are so many other nasty critters out there (worms, Trojan horses, phishing attacks, malicious scripts, spyware, etc.) that the overarching term "Anti-Malware" is more accurate.

Depending on how one chooses to measure, there are from 500,000–40,000,000 malware[2] in the field that impact Android, iOS, macOS, Windows, and the Internet of Things (IoT). Symantec reports they receive as many as 40,000 new signatures in a single day.

---

[1] *https://en.wikipedia.org/wiki/Honoré_de_Balzac*
[2] *https://en.wikipedia.org/wiki/Malware*

iOS presents a locked-down environment, devoid of Flash and Java (and their vulnerabilities), and immune to Windows-specific or macOS-specific malware. With the notable exception of the XcodeGhost debacle[3], to date there few malwares that can impact iOS, and Apple has locked these out with iOS updates. In addition, unless the iOS device has been jailbroken[4], the *only* way to install apps is via the Apple App Store, and Apple has done a very good (not perfect) job validating the integrity of such apps.

On the other hand, if the iOS device *has* been jailbroken, all bets are off. The act of jailbreaking effectively bypasses all that Apple has devised to protect the device. The jailbroken device suffers from an inability to receive operating system updates to close off newly discovered vulnerabilities, and as the purpose of jailbreaking is to install apps from other than the Apple App Store, it is likely malicious or compromised apps are installed.

Should you use anti-malware software on an iOS device? At the moment, the probability of your iOS device being penetrated is remote. However, US-CERT and the NSA warn it is only a matter of *when*, not *if*, a major attack is made on iOS.

As of this writing there are no longer any true anti-malware solutions available for iOS. As soon as a quality anti-malware becomes available for iOS, we announce it on our blog, newsletter, and Facebook page.

As to Android, Chrome OS, macOS, and Windows anti-malware, we recommend Bitdefender. This is based on results in independent testing for catch rate, false positives, resource utilization, ease of use, and cross-platform availability.

8.1.1   [macOS and Windows] Assignment: Install and Configure Bitdefender for Home Users

8.1.2   [macOS and Windows] Assignment: Install and Configure Bitdefender for Business Users

---

[3] *https://en.wikipedia.org/wiki/XcodeGhost*
[4] *https://en.wikipedia.org/wiki/Privilege_escalation#Jailbreaking*

8.1.3 [Windows] Assignment: Disinfect Windows From Outside Windows

8.2 [Android, Chrome OS, macOS, and Windows] Remove Malicious or Problem Applications

8.2.1 [Android, Chrome OS, macOS, and Windows] Assignment: Isolate and Remove Malicious or Problem App

8.2.2 [Android, Chrome OS, macOS, and Windows] Assignment: Protect Your Device from Problem Apps

8.2.3 [Android, Chrome OS, macOS, and Windows] Assignment: Protect Your Device from Problem Apps with Bitdefender

8.3 [Windows] Controlled Folder Access

8.3.1 [Windows] Assignment: Enable Controlled Folder Access

8.4 Advanced Outlook.com Security for Office 365 Subscribers

When using Outlook.com as an Office 365 subscriber, an advanced security module called *Safelinks*[5] is active by default. Safelinks performs extra screening of attachments and links in messages you receive.

---

[5] *https://support.microsoft.com/en-us/office/advanced-outlook-com-security-for-office-365-subscribers-882d2243-eab9-4545-a58a-b36fee4a46e2*

Attachments

As a paid subscriber, *Safelinks* scans all attachments for viruses and malware. If it detects a suspicious file, it is removed so you can't accidentally open it.

Links

When receiving a message with a web link, Safelinks scans to see if the link is related to phishing scams or might download malware. If you do tap a link that is suspicious, you are redirected to a warning page:

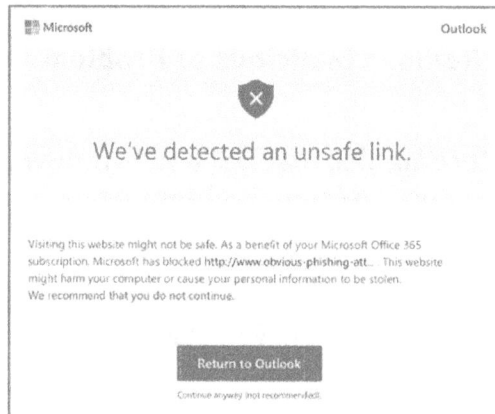

Safelinks Protection

This protection is automatically active with a paid 365 subscription, regardless of how you access your email–from outlook.com or any email client. The protection is only for Outlook.com mailboxes only, and do not apply to third-party accounts such as Gmail that are synched to an Outlook.com account.

8.4.1   [Optional] Assignment: Disable Safelinks in Outlook.com

In this assignment, you disable Safelinks in Outlook.com email.

- Note: This assignment is provided only for informational purposes. We do not recommend disabling Safelinks.

- Prerequisite: A paid Office 365 account with an active Outlook.com email account.

1. Open a browser to *https://outlook.live.com.*

2. Select *Settings > Premium > Security.*

3. Under *Advanced Security,* turn off the switch for *Safelinks.*

- Note: Turning off Safelinks only affects future messages. It does not change the link format in messages already received.

8.5 [Windows] Windows Security Configuration

8.5.1 [Windows] Assignment: Harden Windows Security Configuration

8.6 Lockdown Mode

New to iOS 16 and macOS 13 is *Lockdown Mode.* This brings extreme cybersecurity to iPhone/iPad and Mac users. Although aimed at the small minority of users that may be the subject of highly targeted spyware, we hold that all users should evaluate if such heightened security is appropriate for them.

Once enabled, Lockdown Mode adds the following protections, with Apple promising to continually expand:

- Apple Services
  - Incoming invitations for Apple services from people you haven't previously invited are blocked.
- Browsers using WebKit (including Safari)
  - Some web technologies such as just-in-time (JIT) JavaScript are disabled until the user excludes a trusted site from Lockdown Mode.
- Configuration profiles
  - Profiles cannot be installed.
  - Device cannot enroll into mobile device management (MDM).

- FaceTime
  - Incoming calls from people you have not previously called are blocked.
- Hardware
  - Wired connections with other devices and accessories are blocked when the device locks.
- Messages app
  - Message attachments other than images are blocked.
  - Link previews do not display.
- Photos app
  - Shared albums are removed.
  - New shared album invitations are blocked.

8.6.1 [Optional] Assignment: Enable Lockdown Mode

In this assignment, you enable Lockdown Mode.

1. Open *Settings > Privacy & Security > Lockdown Mode.*
2. Tap *Turn On Lockdown Mode.* The *Lockdown Mode* screen appears.
3. Tap *Turn On Lockdown Mode.* The *Turn on Lockdown Mode?* screen appears.
4. Tap *Turn On & Restart.*
5. Your device restarts with Lockdown Mode enabled.

8.6.2 [Optional] Assignment: Disable Lockdown Mode

In this assignment, you disable Lockdown Mode.

1. Open *Settings > Privacy & Security > Lockdown Mode.*
2. Tap *Turn Off Lockdown Mode.* The *Lockdown Mode* screen appears.
3. Tap *Turn Off Lockdown Mode.* The *Turn off Lockdown Mode?* screen appears.
4. Tap *Turn Off & Restart.*

5. Your device restarts with Lockdown Mode disabled.

8.7   Malware Lessons Learned

☐ Malware is an umbrella term that includes worms, Trojan horses, phishing attacks, malicious scripts, spyware, viruses, etc.

☐ There are from 500,000-40,000,000 malware in the field, with up to 40,000 new signatures appearing daily.

☐ Although Windows-specific malware cannot impact macOS, it can be inadvertently passed along to a Windows device via email or file sharing.

☐ Bitdefender is one of the consistently top-performing anti-malware products. It is cross-platform, inexpensive, and easy to install and configure.

☐ New as of iOS 16 is Lockdown Mode, which adds extreme cybersecurity measures to iPhone and iPad.

8.8   Additional Reading

Souppaya, Murugiah, and Karen Scarfone. "Guide to Malware Incident Prevention and Handling for Desktops and Laptops." NIST Special Publication 800-83, Revision 1. July 2013. <https://nvlpubs.nist.gov/nistpubs/SpecialPublications/NIST.SP.800-83r1.pdf>

9 Firewall

Expecting the world to treat you fairly because you are a good person is a little like expecting the bull not to attack you because you are a vegetarian.

–Dennis Wholey[1]

What You Will Learn in This Chapter

- What a firewall does

What You Will Need in This Chapter

- No additional resources required.

9.1   Firewall

Whenever a computer, mobile device, or network device needs to communicate with the outside world–say, to print, receive or send email, or surf the web–it must *open a door* to that world. In the IT universe, this is called *opening a port*.

Ports are numbered from 1 to 65,535, with at least one unique port number assigned to any one communication task. For example, when using your browser to visit Google, you enter *https://www.google.com* in the address field. This can be translated into English as: *Using the language of the secure Internet (https) I would like to communicate with a server named www, within a domain named google.com.*

---

[1] *https://en.wikipedia.org/wiki/Dennis_Wholey*

The problem is that the www server at Google has 65,535 ports to which it may potentially need to listen. Invisible to the user, *:80* is placed at the end of the address request. This translates into: *And please knock on port 80 (reserved for web server communications) so that www can respond to the web page requests sent to it.*

To best secure your computer, it is important to only have those ports open that are necessary to perform your work.

The purpose of a firewall[2] is to block unwanted attempts to get into or communicate with your computer from the network or Internet through your 65,535 ports. It is about as simple as anything can be on a computer, and once activated you never need to know about it again.

To get into your computer or to communicate with it, the following must be true:

- Your device must be on a network with other devjces (such as your local area network at home or office, or the Internet).

- You must have a port open. On iOS and macOS, ports are opened by enabling sharing services from the Sharing System Setting, and by some applications.

- Lastly, there must be some process or application listening at the port that can respond. You can open port 80 on a web server, but if the web server application (typically Apache) hasn't been launched, no amount of your browser screaming at the server elicits a response.

The odd news is that iOS does not include a firewall. The good news is that there is no need for a firewall with your iPhone or iPad.

As stated in the Apple *iOS Security* whitepaper[3]:

> *On other platforms, firewall software is needed to protect open communication ports against intrusion. Because iOS achieves a reduced attack surface by limiting listening ports and removing unnecessary network utilities such as telnet, shells, or a web server, no additional firewall software is needed on iOS devices.*

---

[2] *https://en.wikipedia.org/wiki/Firewall*
[3] *https://www.apple.com/business/docs/iOS_Security_Guide.pdf*

9.1.1 [macOS and Windows] Assignment: Activate the Firewall

9.1.2 [macOS and Windows] Assignment: Close Unnecessary Ports

9.2 Firewall Lessons Learned

☐ Other devices and services on the local network and internet communicate with your device through logical ports, numbered 1 to 65,535, each with a unique function.

☐ To best secure a device, only necessary ports should be open.

☐ A firewall blocks unwanted attempts to communicate with your device.

☐ iOS and iPadOS do not have a firewall and have no need for a firewall.

10 Lost or Stolen Device

It takes considerable knowledge just to realize the extent of your own ignorance.

–Thomas Sowell[1], American economist, social theorist, political philosopher, and currently Senior Fellow at the Hoover Institution, Stanford University[2]

What You Will Learn in This Chapter

- Activate and configure Find My iPhone/iPad
- Use Find My from a computer
- Use Find My from an iPhone or iPad

What You Will Need in This Chapter

- No additional resources required.

10.1   Find My iPhone

Millions of iPhones and iPads are stolen each year. If you have followed the steps above to enable strong login password, or enable erasure with 10 unsuccessful password attempts, and set Screen Lock to a very short time frame, all a thief gets is the device–not the data.

But it would be nice to be able to get your device back.

---

[1] *https://en.wikipedia.org/wiki/Thomas_Sowell*
[2] *This observation has since been validated in University studies, originally performed by David Dunning and Justin Kruger. It is now referred to as the Dunning-Kruger effect. https://en.wikipedia.org/wiki/Dunning–Kruger_effect*

Find My iPhone is an option within iCloud accounts that locates your iOS device on web map, often to within 6 feet. By passing this information along to your local law enforcement, they can get a search warrant to the address and (possibly) recover your property. Or as in my all-too-often case, dig down in the back yard to find where the Yellow Lab has buried the device.

Your missing/stolen device occasionally sends out a Bluetooth beacon that is picked up from nearby Apple devices (Mac, iPhone, and iPad). These devices know their location. They relay their location and presence of your device to Apple. When you open the Find My app on another of your Apple devices, or log into icloud.com, the location of your missing device is displayed on a map.

For Find My to function, the following must happen:

- An iCloud account has been activated.
- *Find My* has been enabled for the device.
- The device is turned on, even if locked.
- Another Apple device is active and within Bluetooth range.

10.1.1   Assignment: Activate and Configure Find My iPhone/iPad

In this assignment, you configure your iOS device to communicate its location to your iCloud account.

- Prerequisite: iCloud account is active on this device.

1. Select *Settings > Your Name,* then tap *Find My iPhone*.

2. In the *Find My iPhone* screen, enable *Find My iPhone, Find My network,* and *Send Last Location.* This option shows the last seen location in the event the battery or device dies.

3. Exit *Settings*.

Your iOS device is now continuously broadcasting its GPS location data to your iCloud account.

10.1.2   Assignment: Find Your Device from a Computer

For the purposes of this assignment, let's assume you have lost, or someone has taken, your iPhone or iPad (or perhaps like me, you have a slightly neurotic Yellow Lab that is obsessed with burying the device).

In this assignment, you find the device using any computer.

1. From a computer, open a browser, go to *https://www.icloud.com*, then enter your Apple ID and password.

2. The iCloud desktop appears. Select the *Find iPhone* button.

3. The *Find My iPhone* map appears. All your registered devices appear. If the device is powered on, it has a green light next to it. If the iOS device *Settings > iCloud > Send Last Location* is set to *On*, even if the device is off or destroyed, its last transmitting location is marked.

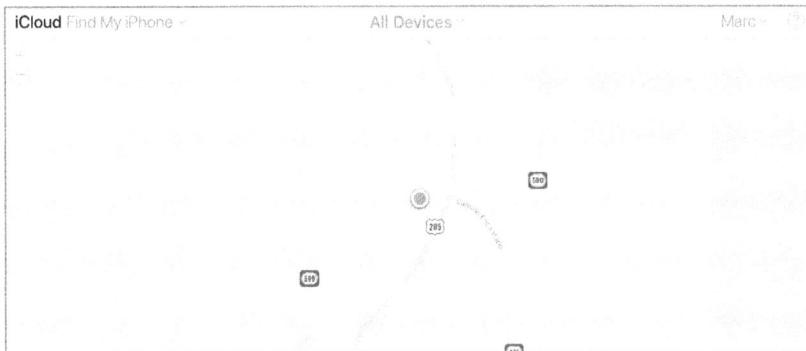

4. Taping *All* Devices at the top center of the window displays all your devices.

5. Select a device from the *My Devices* list, highlighting it on the map.

- *Play Sound.* Plays a "sonar" sound and display an alert window on the device. This is most useful when Rover has decided to bury your device along with all his other most precious possessions. I do not recommend this option if the device has been stolen, as it notifies the thief that you are tracking them, perhaps prompting them to destroy the "tracking device."

- *Lost Mode.* Locks down your device, preventing any use. This is most useful if the device has been stolen.

- *Erase iPhone* or *Mac.* If it is not possible to get prompt police intervention, you have valuable data on the device, and the thief may have access to a server farm to perform a massive password hack attempt, you may want to consider simply erasing your device. After all, you do have a full current backup on iCloud and computer, don't you?

10.1.3   Assignment: Find Your Device from an iOS Device

For this assignment, you find your device using an iPhone or iPad.

1. Install the *Find My iPhone* app from the App Store.

2. Open the *Find iPhone* app.

3. At the prompt, log into your iCloud account.

4. Open the *Find My iPhone* app, then tap *Devices* at the bottom center. All your devices registered with *iCloud* display. Tap on the target device to locate.

5. Tap on the lost device name. The map opens, displaying the exact last known location for the target device.

6. Below the map you see the *Actions* screen. From here you may have the device play a sonar sound (I use this at least once a week–my Yellow Lab is obsessed with hiding my iPhone), lock it to prevent any access, or erase the device.

7. When done, exit the *Find My* app.

Hooray! You've found your lost/stolen device.

10.2 [Android and Windows] Bitdefender Anti-Theft

10.2.1 [Android, Chrome, and Windows] Assignment: Activate and Configure Bitdefender Anti-Theft

10.2.2 [Android, Chrome, and Windows] Assignment: Find a Device from a Computer

10.3 Lost or Stolen Device Lessons Learned

☐ *Find My iPhone* is built into iOS. Once enabled, it is possible to locate your device if powered on.

☐ Enable Find My iPhone from *Settings > Your Name*, then tap *Find My iPhone*.

☐ Bitdefender *Anti-Theft* does not work with iOS.

11 Local Network and Bluetooth

I am concerned for the security of our great Nation; not so much because of any threat from without, but because of the insidious forces working from within.

–General Douglas MacArthur[1]

What You Will Learn in This Chapter

- Block Ethernet broadcasting
- Prevent Ethernet insertion
- Understand Wi-Fi encryption protocols
- Configure router security
- Secure Bluetooth devices

What You Will Need in This Chapter

- No additional resources required.

11.1 Ethernet Broadcasting

It is common wisdom that Ethernet is more secure than Wi-Fi. But as with most things we believe, this is not accurate.

There are two security issues with Ethernet: Broadcasting and Insertion. At the most fundamental level, what happens when data travels through Ethernet is that electrons travel along a metal cable. There are two unintended consequences that

---

[1] *https://en.wikipedia.org/wiki/Douglas_MacArthur*

occur whenever electrons go for a ride: heat generation, and creation of an electromagnetic field. For our purposes, heat is not an issue. But the electromagnetic field is.

Sending data through copper wire effectively turns that wire into a very large antenna that broadcasts your data through radio waves. With the right receiver and translation software, you can easily capture every bit of data being sent and received along that cable.

This vulnerability is not something about which the average person or business would or should be concerned. On the other hand, if you or your business requires the utmost in security, it is mandatory to add encryption to your Ethernet network. This can be done by using a *Virtual Private Network* (*VPN*) service or installing a RADIUS server on your network. VPN is discussed in the Internet chapter. RADIUS use is beyond the scope of this book.

11.2 Ethernet Insertion

You would notice if someone came into your home, plugged a computer into your network, and sat there watching data go by. But in a typical business, nobody would notice.

Ethernet and Wi-Fi networks can be protected from unwanted insertions by implementing the 802.1x protocol[2] (often referred to as RADIUS). This protocol works with both Ethernet and Wi-Fi, mandating that anyone attempting to join the network authenticate with their own personal name and password. This is unlike the typical Wi-Fi authentication that uses the same password for everyone.

To implement 802.1x you need to have either: a macOS, Windows, or Linux Server running within your network; one of the many other 802.1x appliances that are sold; or even a cloud service that provides RADIUS service[3]. Details on how to configure 802.1x are beyond the scope of this book.

---

[2] *https://en.wikipedia.org/wiki/IEEE_802.1X*
[3] *https://jumpcloud.com/blog/mac-radius-authentication*

11.3   Wi-Fi Encryption Protocols

Right out of the box, many Wi-Fi base stations are insecure. Anyone who can pick up the signal can connect. This allows them to not only use your bandwidth to access the Internet, but also to see all the other data such as usernames and passwords that travel on that network. To start securing your Wi-Fi, add strong password protection with encryption.

Although cellular networks do use encryption, the protocol in use has been broken for many years, making it easy for a novice hacker to see all data passing through. In addition, it is common practice for police and other government agencies to set up their own cellular towers[4] (Stingray) with the purpose of harvesting data.

To prevent your data from being seen while on a cellular network or an insecure Wi-Fi network, it is necessary to use VPN (Virtual Private Network) encryption (more on that later). If the Wi-Fi network is properly encrypted, you should have little concern about the security and privacy of your data.

Below is a brief on each of the Wi-Fi encryption protocols.

- **WEP**[5] (Wired Equivalency Protocol) was the first encryption protocol for Wi-Fi. Introduced in 1999, it was quickly broken, and by 2003 was replaced by WPA and WPA2 (Wi-Fi Protected Access). Any Wi-Fi base station manufactured in the past 5 years offers WPA and WPA2 in addition to WEP.

There is only one reason to ever use WEP: you simply have no other option. Kids driving by your home can break into your WEP network before leaving the block.

- **WPA**[6] (Wi-Fi Protected Access) superseded WEP in 2003. Although it is a great advancement, it too has been broken. As with WEP, the only reason to use WPA is that you have no other option.

- **WPA2**[7] superseded WPA in 2004. Although technically difficult, WPA2 can also be hacked.

---

[4] *https://en.wikipedia.org/wiki/Stingray_phone_tracker*
[5] *https://en.wikipedia.org/wiki/Wired_Equivalent_Privacy*
[6] *https://en.wikipedia.org/wiki/Wi-Fi_Protected_Access*
[7] *https://en.wikipedia.org/wiki/Wi-Fi_Protected_Access*

- **WPA3** Superseded WPA2 with certification beginning June 2018. All wi-fi devices certified for WPA after June 30, 2020, must include support for WPA3.

Two encryption algorithms are available to you: *TKIP* and *AES* (technically known as CCMP, but virtually all vendors refer to it as AES). TKIP has been compromised and is no longer recommended. If your Wi-Fi device allows the option of AES, use only that. If it only allows for TKIP, trash the unit, then purchase a more modern device.

11.4 Network Device Overview

The connection point between your Internet Service Provider (ISP) and your Local Area Network (LAN) is a router. A router is a device designed to connect two different types of networks and provide resources for them to interact.

Common brands of routers include Cisco, Ubiquiti, Linksys, Netgear, D-Link, and many unbranded devices that Internet Service Providers lease to customers.

Some newer routers, especially those provided by ISPs, are all-in-one units containing several if not all the components below:

- **Modem**[8]. This hardware decodes and modulates the signal from your Internet provider to your cable or telephone jack. A modem is likely to be a separate component if more than one device exists for your Internet connection.

- **Router**[9]. This component runs a specialized program that allows hundreds of different devices to interact on a network, usually sharing a single IP address to the Internet. Routers use *Network Address Translation* (NAT) to convert and direct Internet traffic from websites to your computer and from your computer to other computers and peripherals on the *Local Area Network* (LAN).

---

[8] *https://en.wikipedia.org/wiki/Modem*
[9] *https://en.wikipedia.org/wiki/Router_(computing)*

- **Firewall**[10]. A firewall may be either software or hardware. Most modern routers now include a firewall, as do most operating systems. The firewall inspects data traffic between the internet and internally connected devices.

- **Intrusion Detection System** and **Intrusion Prevention System**[11]. These features are built into high-end routers. They perform deep packet inspection to further protect the network from outside intruders.

- **Network Switch**[12]. This hardware component allows multiple devices to be connected simultaneously and interact with the router using Ethernet cables. A switch is the modern version of an ethernet hub. Hubs present a critical security vulnerability and should never be used.

- **Access Point**[13]. An access point hardware component allows tens or hundreds of wireless (Wi-Fi) devices to connect to it.

Every router has at least some basic security controls built in. These controls include the ability to filter out what the router thinks are attempts to hack into your network, and the ability to forward specific types of data packets to a specific computer in your LAN, or to point specific types of data packets to a specific computer on the Internet.

Malware, hackers, criminals, and even some government agencies, sometimes attempt to alter security control configurations so that the malware or perpetrators have an easier time harvesting your data. Because of this, it is wise to routinely inspect the condition of your router. How often is *routine?* Within larger or security-conscious organizations with high-value data, it is common to have a network administrator dedicated to maintaining watch over the status of network equipment. For a small business or household, once a month is not too often.

11.4.1   Assignment: Determine Your Wi-Fi Encryption Protocol

---

[10] *https://en.wikipedia.org/wiki/Firewall_(computing)*

[11] *https://en.wikipedia.org/wiki/Deep_packet_inspection*

[12] *https://en.wikipedia.org/wiki/Network_switch*

[13] *https://en.wikipedia.org/wiki/Wireless_access_point*

It is vital to know if the Wi-Fi network your device connects to is secure with WPA2 or WPA3. If it is not, almost *everything* your device does on that network may be viewed-including usernames and passwords.

Unfortunately, iOS 16 does not include the tools to see what encryption protocol is in use. However, you can see if there is encryption in place, and test to find which protocol is in use.

In this assignment, you determine the Wi-Fi encryption protocol that is in use.

- Prerequisite: An available Wi-Fi network for which you know the password.

1. Open *Settings > Wi-Fi*. If you see a lock icon to the right of the desired network, that network has encryption. The bad news is there is no way to know if it is WEP, WPA, WPA2, or WPA3.

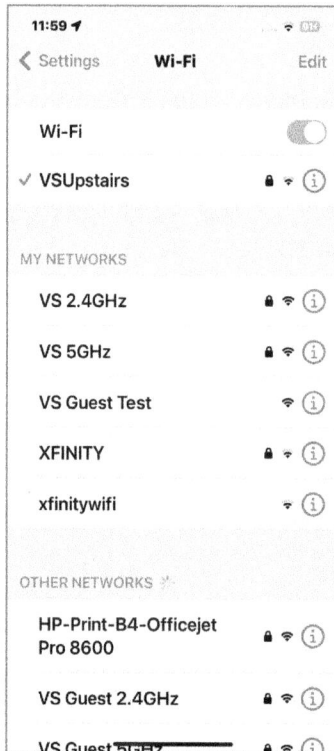

2. Scroll to the bottom of the screen, then tap *Other…*

3. In the *Other Network* screen, enter the name of an available Wi-Fi network, enter the password, then set the *Security* to either *None, WEP, WPA, WPA2, WPA Enterprise,* or *WPA2 Enterprise,* (if WPA Enterprise or WPA2 Enterprise is selected enter your username), then tap *Join.*

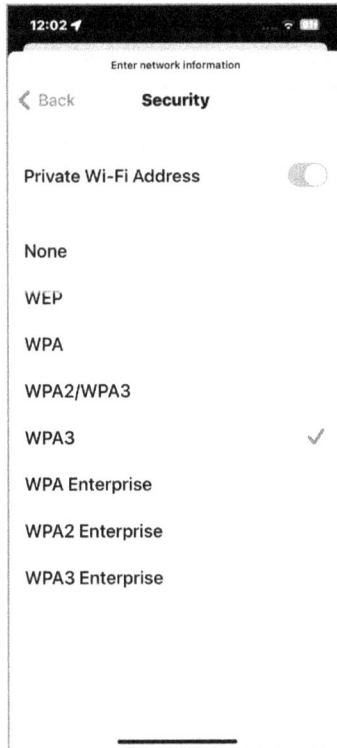

4. If you can join the network, then the choice of encryption is correct. If you are not able to join the network, repeat step 3 using a different encryption protocol until you are able to join.

11.4.2 Assignment: Configure WPA2 or WPA3 on Your Router

This assignment is generic, as each router has its own unique interface to configure Wi-Fi encryption.

In this assignment, you configure your Wi-Fi router to use the secure WPA2 or WPA3 protocol.

- Note: If this assignment is performed in a class, the instructor demonstrates while the students observe.

- Prerequisite: You need the administrator's name and password of the router to view and edit the current settings, and to change settings.

- Prerequisite: You need the user manual for your router.

Find the IP address of your Wi-Fi router

1. Open *Settings.app* > *Wi-Fi* > tap the *Info* icon for your current Wi-Fi network, then scroll down to *Router.* The router address displays.

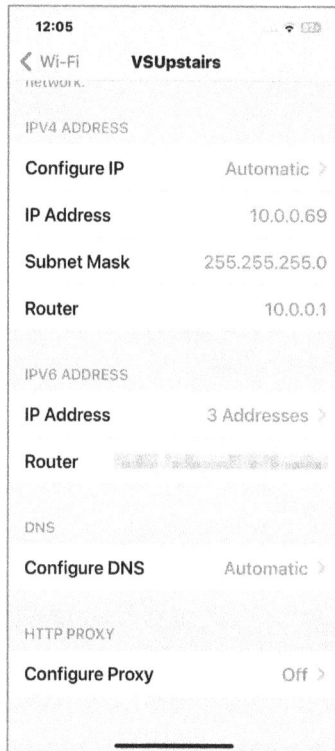

12:05	🔋
‹ Wi-Fi **VSUpstairs**	
network.	
IPV4 ADDRESS	
Configure IP	Automatic ›
IP Address	10.0.0.69
Subnet Mask	255.255.255.0
Router	10.0.0.1
IPV6 ADDRESS	
IP Address	3 Addresses ›
Router	
DNS	
Configure DNS	Automatic ›
HTTP PROXY	
Configure Proxy	Off ›

2. Exit *Settings.app*.

Access the router

3. Open a web browser. Then in the URL or address field, enter the IP address of the Wi-Fi base station or router.

4. At the *Authentication* window, enter the administrator's name and password. This is the administrator of the router, not of your device.

5. The router control panel appears.

● Note: Below is a screenshot of an ASUS GT-AXE11000 router. Each router has its own unique interface.

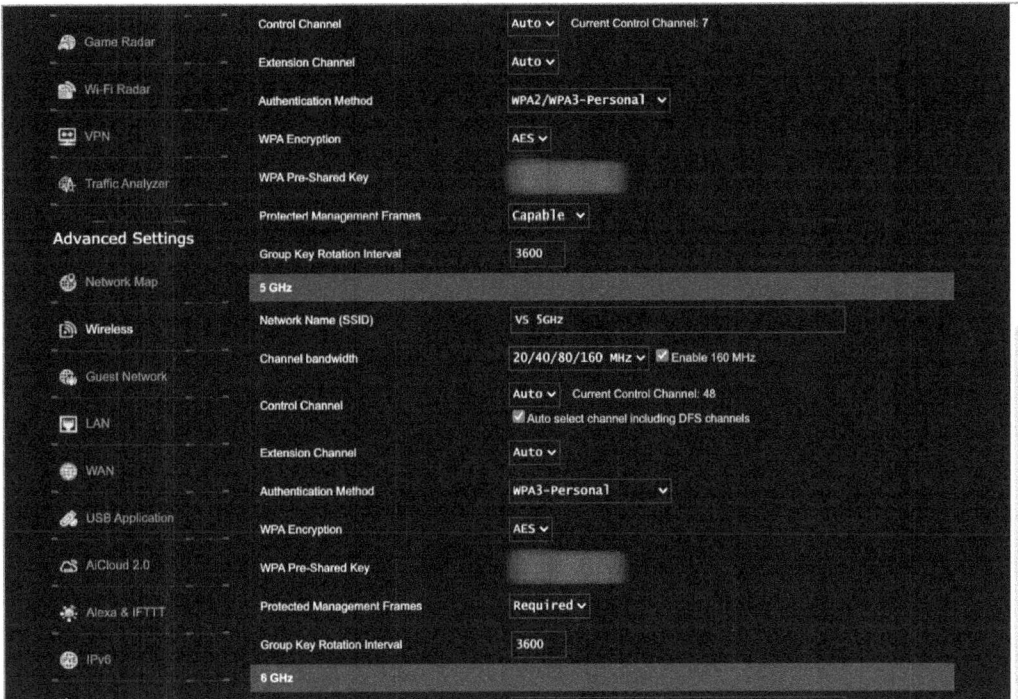

6. Follow the instructions in the router manual to configure Wi-Fi encryption to WPA2, or if available, WPA3.

● Note: If your router has the option of using either *AES* or *TKIP*, select *AES*. If you only have an option for TKIP, immediately replace this router. The TKIP encryption scheme has been broken and is easily hacked.

7. If any changes were made, tap the *Apply* button to save the changes.

8. Close the browser window to exit out of your router.

Congratulations! All traffic on your Wi-Fi is now securely encrypted.

11.5 Use MAC Address To Limit Wi-Fi Access

Every device that can connect to a TCP network has a unique *MAC Address*[14] (Media Access Control), sometimes referred to as the *Hardware Address*. This address specifies the manufacturer of the device, and a device-specific number. Do not go to sleep on me yet! This MAC address can be used with most Wi-Fi base stations to limit what devices can connect to your network.

Although every Wi-Fi base station has a unique interface to filter by MAC address, they all operate on the same principle: either allow anyone with the proper password to gain access to the network or allow anyone with the proper password *and* proper MAC address access to the network. In this way, you can easily lock down your Wi-Fi to only approved devices. So even if employees know the password, they are unable to connect their personal device to the Wi-Fi unless the MAC address for those devices is on the list.

- Note: Consider MAC Address Filtering as something to keep the kids out of the network, not skilled hackers. It is possible to see the MAC addresses of some network-connected devices, then to temporarily fake having one of those MAC addresses, which can help a hacker get onto the network.

11.5.1 Assignment: Restrict Access by MAC Address

In this assignment, you configure a router to allow only desired devices to connect. This assignment is generic, as each router has its own unique interface.

- Note: If this assignment is performed in a class, the instructor demonstrates while the students observe.

- Prerequisite: You need the administrator's name and password of the router to view and edit the current settings.

- Prerequisite: You need the user manual for your router.

Find and record the IP address of your permitted wireless devices

---

[14] *https://en.wikipedia.org/wiki/MAC_address*

1. Make a list of devices to be permitted access to your Wi-Fi network. Include a 1 or 2-word description, and the MAC address of the device.

Device Name	User	Description	MAC
Roku TV	Staff	58" Conf room	XX:XX:XX:XX:XX
MacBook	Marc	2020 13" M1	XX:XX:XX:XX:XX

- The MAC address of an Android device may be found from *Settings > Connections >* tap your *current network name >* tap *Gear* icon next to the current network name > *Advanced >* scroll to the bottom of screen for *MAC address.*
  - Note: The *MAC address type* must be set to *Phone MAC,* not *Randomized MAC,* or it constantly changes.

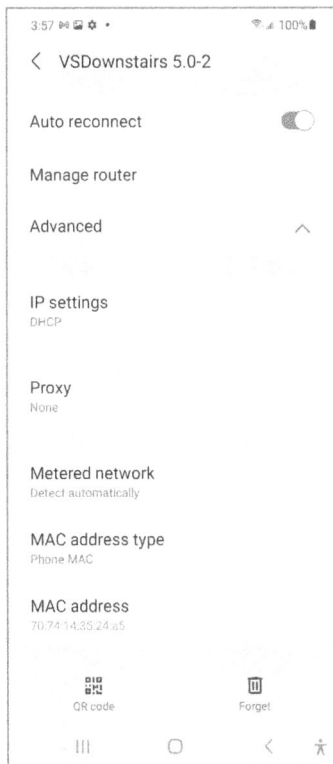

- The MAC address of a Chrome OS devices may be found from *Settings > Network > current network name > current network name > expand Network area > MAC address* field.

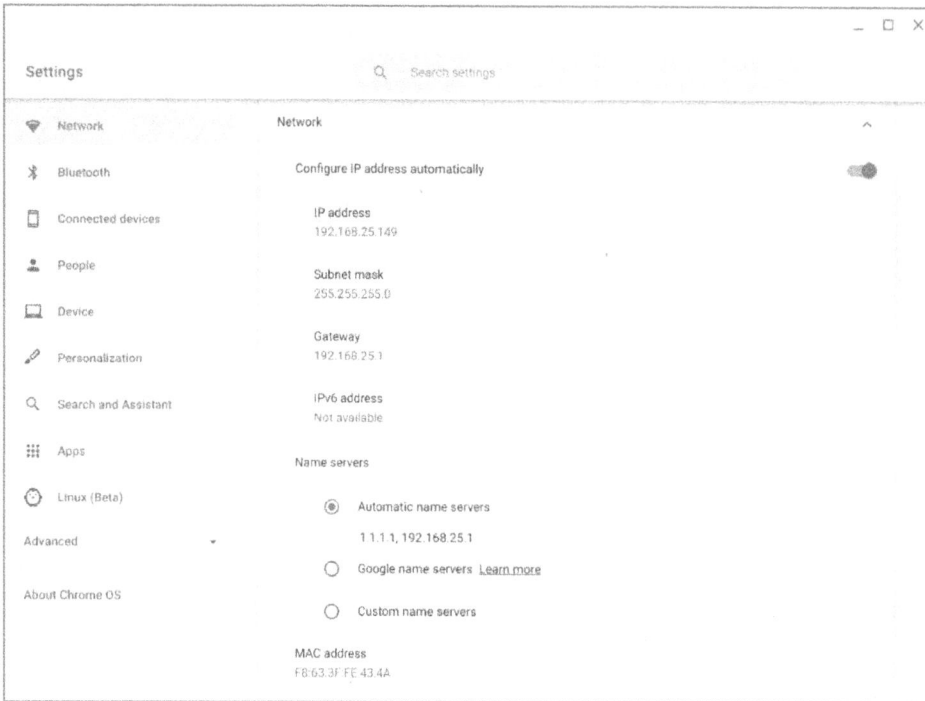

- The MAC address of an iPhone and iPad is found in the *Settings > General > About > Wi-Fi Address* field.

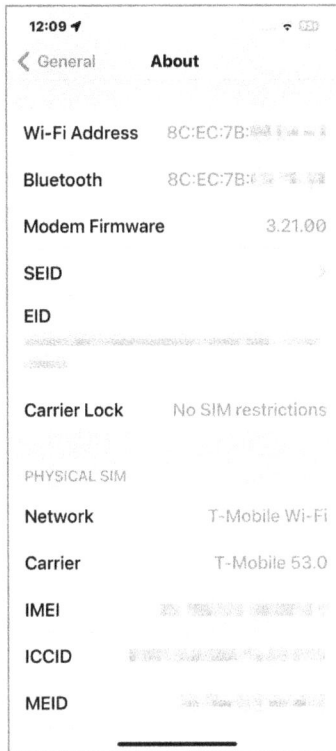

- The MAC address of a macOS 13 device may be found from the *Apple* menu > *System Settings* > *Network* > connected network > *Details* button > *Hardware* > *MAC address*:

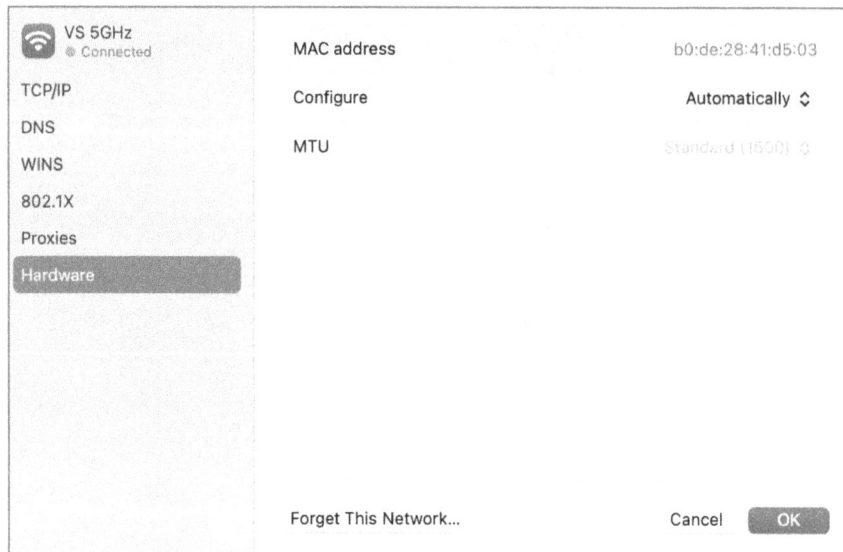

- The MAC address of a Windows 11 device is found with the *ipconfig* command in the command prompt.

  a. In Windows 11, tap the *Search the Web and Windows* field in the bottom left corner, then enter command prompt. Double-tap on Command Prompt in the *Best match* pop-up.

  b. The Command Prompt window appears.

  c. Enter `ipconfig -all`. A listing of all network addresses for the device appears. The MAC address shows as the *Physical Address*.

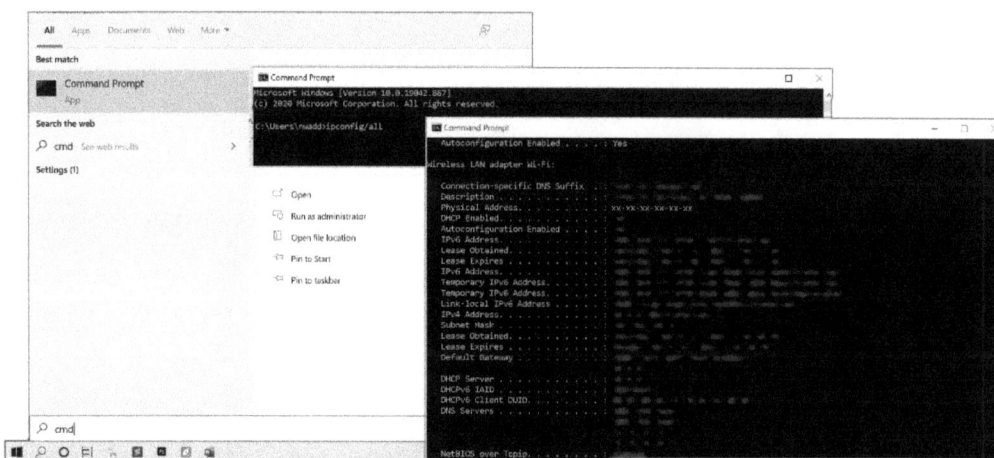

    d.   Close the Command Prompt.

Configure the router

2. With your list of desired devices in hand, launch a browser and enter the IP address of your Wi-Fi router.

3. At the authentication window, enter the name and password of the administrator for the router.

4. The router control panel appears.

   - Keep in mind that all routers–even from the same company–have slightly different interfaces.

5. Follow the user manual for your router to enter the page for *MAC Filter,* then enter the MAC address and device description for each device to be allowed Wi-Fi access to your network.

6. *Save* changes.

7. Close the browser window to exit out of your wireless router.

Congratulations! You have secured your wireless network so that only authorized devices have access.

11.6 Router Penetration

The connection point between your Internet provider cable, DSL, fiber, radio, etc. and your Local Area Network (LAN) is a router. A router is a device designed to connect two different types of networks.

Common areas of router penetration include:

- **Port forwarding**[15]**:** Port forwarding allows a service–such as an internal web server–to be accessible from the internet. However, if ports are being forwarded without purpose, the firewall is being bypassed and your internal computers may be visible from the internet.

---

[15] *https://en.wikipedia.org/wiki/Port_forwarding*

- **DMZ**[16]**:** Related to Port Forwarding is the DMZ, or De-Militarized Zone. DMZ is typically used to route *all* external traffic for a specific IP address to a specific computer, regardless of service request. Unless there is a unique need, it should remain disabled or else it may be used for nefarious purposes.

- **RAM-Resident Malware:** Some router malware make their home in the RAM of the router. In this way, they can take control of your data traffic without showing in the interface.

- **Firmware**[17]**:** It is vital to keep the router firmware up to date. Just as with any software, router firmware always has vulnerabilities. Over time, criminals (and some government organizations) discover how to use these vulnerabilities to their benefit. Keep the firmware updated to stay a step ahead of this problem.

11.6.1 Assignment: Verify Router Security Configuration

In the example below, I use an ASUS GT-AXE11000. Although all routers have a different interface, most share the same functions.

In this assignment, you verify the security configuration of a Router.

- Note: If this assignment is performed in a class, the instructor demonstrates while the students observe.

- Prerequisite: You need the administrator's name and password for your router.

- Prerequisite: You need the user manual for your router.

Remove RAM-resident malware

Some malware make their home in the router RAM. Also, over time the router RAM may accumulate corruption. The fix for both issues is the same: power cycling.

1. Verify that all users have disconnected from the network and Internet and have closed any connections to other devices on the network.

---

[16] *https://en.wikipedia.org/wiki/DMZ_(computing)*
[17] *https://en.wikipedia.org/wiki/Firmware*

2. Power off the router. If your router does not have an on/off switch, pull the power cord from the back of the router.

3. Remove the router batteries (if any).

4. Wait a minute.

5. Insert the router batteries (if any).

6. Power on the router. It may take up to 5 minutes for it to be fully operational.

7. Open a browser and enter the IP address of your router.

8. At the prompt, enter the administrator's name and password.

Verify router firmware is up to date

9. Follow your router user manual to check if your firmware version is current. If not, follow your router user manual to update the router firmware. The typical steps are to:

   a. From a browser logged into the router control panel, download the current firmware.

   b. From the router control panel, upload the firmware to the router.

   c. From the router control panel, install the new firmware.

   d. Restart the router.

Verify no unnecessary port forwarding

10. Follow your router manual to display the *Port Forwarding* page.

11. Verify Port Forwarding is disabled. If it is enabled, verify the need for the activity.

Verify DMZ configuration

DMZ is like *Port Forwarding*. When *DMZ* is enabled, all inbound packets are routed to the specified device. This allows a single device on your network to be accessible from the Internet. Such access presents a high level of vulnerability for that device. Unless there is a demonstrated business need for this function, and adequate steps have been taken to prevent unwanted penetration, turn off *DMZ*.

12. Follow your router manual to display the DMZ page.

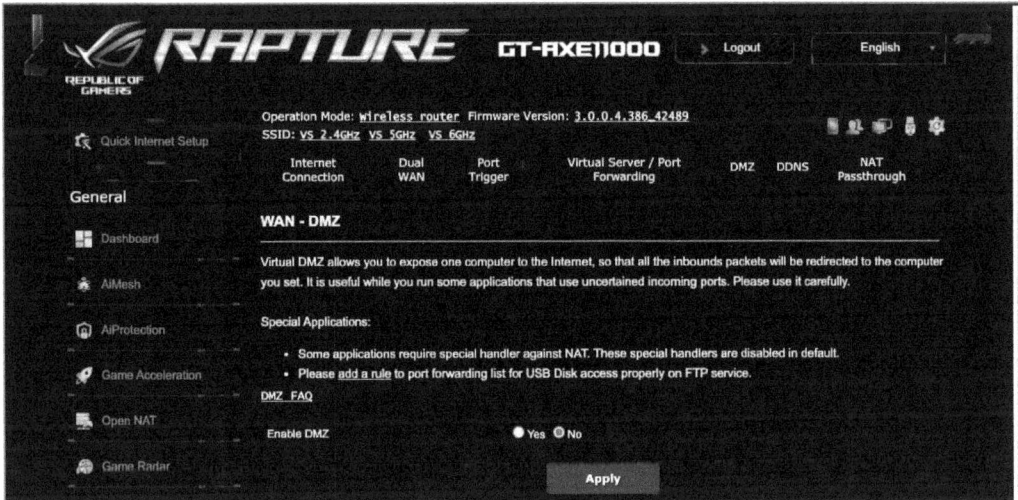

13. Verify *DMZ* is disabled. If it is enabled, verify the need for the activity and disable DMZ if there is no need. In the very rare environment where enabling DMZ is required, understand that all inbound traffic is forwarded to the target device, and that device loses all protection of the router's firewall. The device is fully exposed to the internet and must include tools to protect itself from attack.

14. Exit the browser.

Congratulations, your router is in great shape. Remember to perform this same checkup at least monthly.

11.7   Bluetooth

Bluetooth[18] is a wireless standard very similar to Wi-Fi but intended for short range (typically under 30 feet), and for use with peripheral devices with low bandwidth needs such as mice and keyboards. It uses the 2.4 GHz spectrum, which is shared with older Wi-Fi, microwave ovens, garage door openers, wireless headphones, and wireless landline phones.

---

[18] *https://en.wikipedia.org/wiki/Bluetooth*

All modern computers, mobile devices, and Bluetooth devices require authorization prior to linking (also called pairing) one device to the other. This is usually in the form of entering a unique code on one or both devices.

As Bluetooth devices are typically designed to provide some control over the computer – for example, by using a mouse or keyboard – there is significant vulnerability in allowing an unknown Bluetooth device access to your system. Preventing unapproved access is key to maintaining Bluetooth device security.

11.7.1   Assignment: View Paired Bluetooth Devices

To verify the devices connected to your iPhone or iPad via Bluetooth:

1. On your iOS device, open *Settings.app > Bluetooth*.

2. All connected devices display under *My Devices*.

11.7.2   [Optional] Assignment: Add a Bluetooth Device

This is the typical process to connect a Bluetooth device to an iPhone or iPad.

*   Prerequisite: A Bluetooth device to connect to your device.

*   **Warning**: Never authorize a Bluetooth connection unless you are trying to pair it for the first time. If an authorization notice appears and you are not attempting to pair a device, it may be a bad actor attempting to connect to your computer.

1. To prevent unauthorized access to your Bluetooth device or computer, adding a Bluetooth device should be performed in an environment away from others– preferably at least 60 feet (although Bluetooth can be manipulated to transmit over one kilometer.

2. To prevent unauthorized access to your computer, do not approve a Bluetooth pairing request unless you are in the process of adding a device.

3. On your iOS device, open *Settings.app > Bluetooth*.

25. If Bluetooth is off, turn it on.

26. Power on the target Bluetooth device.

27. Within a few seconds, the Bluetooth device appears in your iOS device Bluetooth screen.

28. Tap *Pair or* enter the displayed code.

29. You are now paired with the Bluetooth device. The pairing remains intact until one of the devices is powered off, separated by more than approximately 30', or you *Forget This Device.*

11.7.3 [Optional] Assignment: Remove a Bluetooth Device

Should you no longer need to use a Bluetooth device, it has been lost, or you find an unrecognized device on your list, remove it from your paired device list.

1. On your iOS device, open *Settings.app > Bluetooth.*

2. To the right of the target Bluetooth device, tap the *Info* icon.

3. Tap *Forget This Device.*

4. At the confirmation dialog, tap *OK.*

11.7.4 Assignment: Security Protocols for Bluetooth

Listed below are the NSA-recommended guidelines[19] for working with Bluetooth devices.

- Choose a device that provides only the features needed to accomplish its purpose, without including extraneous functionality that could be exploited by an attacker who wants access to your device or data.

- Pair the device in a secure area, away from possible eavesdroppers.

- If possible, disable profiles/services and features that are not being actively used.

---

[19] *https://apps.nsa.gov/iaarchive/customcf/openAttachment.cfm?FilePath= /iad/library/ia-guidance/tech-briefs/assets/public/upload/Bluetooth-for-Unclassified-Use-Guidelines-for-Users.pdf&WpKes=aF6woL7fQp3dJirHamRT5DfsLHqKp4xf9Q9ATR*

- Do not authorize connections from new devices or enter/approve a passkey unless you are trying to pair devices for the first time. [You could link to the enemy.]

- Choose a device that clearly communicates its status through audio, text, graphics, and/or LEDs.

- Always maintain physical control over your device. If you lose a device, remove it as soon as possible from the paired device list.

- If applicable to the device, apply patches regularly, use device firewalls, and keep anti-virus software up to date.

- If working in an enterprise setting, comply with all Bluetooth policies and guidance for that enterprise.

11.8 Local Network and Bluetooth Lessons Learned

☐ Ethernet broadcasting describes how data passing through ethernet cabling emits an electromagnetic field that can be read from a distance. This is a reason why all traffic in and out of your devices should be encrypted.

☐ Ethernet insertion is when an unauthorized device is connected to your ethernet network. This is a reason why only necessary ethernet ports should be active.

☐ RADIUS (the 802.1x protocol) is a strategy to encrypt all ethernet and Wi-Fi traffic on your network and devices.

☐ WEP is a legacy (old) Wi-Fi encryption protocol that should not be used.

☐ WPA is a modern Wi-Fi encryption protocol, currently at WPA3. WPA2 and WPA3 are the only versions to be used because the original WPA is no longer secure.

☐ A modem is a hardware device the encodes and decodes and modulates the signal between your internet provider and your network.

☐ A router allows multiple devices to interact on a local network and internet.

☐ A firewall is software or hardware that inspects data traffic between the internet and a device.

☐ Intrusion Detection System and Intrusion Prevention System are features that perform deep packet inspection to protect a network from intruders.

☐ Network Switch is a hardware component that allows multiple devices to be connected simultaneously with a router.

☐ Access Point is a hardware device (for example, a wireless router) that allows multiple Wi-Fi devices to connect with the network.

☐ Any device that can connect to a network or Internet has a Media Access Control (MAC) address which provides a device-specific number.

☐ A router can be compromised with unauthorized port forwarding, sending data to a device without your permission. Disable unauthorized port forwarding

☐ A router can be compromised with unauthorized DMZ, sending all traffic to a device that you have not specified. Turn off DMZ.

☐ A router can be compromised with RAM-resident malware. Such malware can be removed by power-cycling the router.

☐ All routers have firmware that requires updating from time to time. See your user manual for instructions.

☐ Bluetooth is a wireless standard very similar to Wi-Fi, but intended for short range, typically under 30 feet.

☐ Bluetooth uses the same unregulated 2.4 GHz frequency range as do older Wi-Fi, microwave ovens, garage door openers, wireless headphones, and wireless landline phones.

☐ Bluetooth devices can be paired with your computer from *System Settings > Bluetooth.* Pair only the devices you recognize.

12 Web Browsing

Distrust and caution are the parents of security.

–Benjamin Franklin[1]

What You Will Learn in This Chapter

- HTTPS good, HTTP not
- Choose a browser
- Configure browser security
- Private browsing
- Secure web searches
- Clear browser history
- Find, remove, and add browser extensions
- Detect fraudulent websites
- Do Not Track and Fingerprinting
- Recover from a web scam
- Use Tor
- Surface web, deep web, and dark web
- Find and fix hacked accounts
- Manage ad targeting information

What You Will Need in This Chapter

---

[1] *https://en.wikipedia.org/wiki/Benjamin_Franklin*

- No additional resources required.

12.1 HTTPS

Due to an extraordinary marketing campaign, everyone knows the catchphrase: *What happens in Vegas, stays in Vegas*. With few exceptions, web surfers think the same thing about their visits to the web.

As of this writing, over 30% of all websites[2] still use the very vulnerable HTTP[3] (Hypertext Transport Protocol) to relay information and requests between user and website and back again. HTTP sends all data in clear text. This means anyone snooping on your network connection anywhere between your computer and the web server can easily see everything you are doing.

The primary alternative to HTTP is HTTPS[4] (Hypertext Transport Protocol Secure). HTTPS accounts for 68% of active websites and uses the SSL[5] (Secure Socket Layer) encryption protocol to ensure all traffic between the user and server is military grade encrypted.

- Note: Due to the continuing efforts of the major web browsers to eradicate insecure sites from the web, at this time, a user of Firefox, Chrome, etc. must set special security exemptions in order access data from an insecure site. Also, further restrictions in Google Search Engine Optimization[6] (SEO) guidelines have removed nearly all insecure protocol website listings from their index. This does leave one to wonder why we have not yet reached universal encryption.

---

[2] *https://w3techs.com/technologies/details/ce-httpsdefault*
[3] *https://en.wikipedia.org/wiki/Hypertext_Transfer_Protocol*
[4] *https://en.wikipedia.org/wiki/HTTPS*
[5] *https://en.wikipedia.org/wiki/Transport_Layer_Security*
[6] *https://en.wikipedia.org/wiki/Search_engine_optimization*

Any time you visit a web page that is secured using https, it is reflected in the URL or address field of your web browser, typically as a small lock icon to the left of the URL. It looks like this:

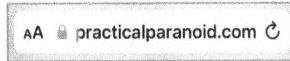

When connected securely to a website, snoops cannot monitor your actions. However, they still can see you are connected to the site. If you want to shield yourself completely, continue reading to our chapter on using a Virtual Private Network (VPN.)

Fortunately, most (but not all) websites that use HTTPS automatically route a browser to the websites' HTTPS even if a browser requests a non-secure HTTP format.

Having to remember to connect only to an HTTPS site for each web page is an impossible task. As we noted, over 30% of websites still do not have an HTTPS option, resulting in an error page and wasted time when you presume to add an S to an HTTP address. But whenever possible you do prefer to work with HTTPS sites because otherwise all your communication is visible to people you do not know/authorize.

There are three options to resolve this:

- Use Safari in iOS 15 or higher. Safari in iOS 15 and higher has a built-in mechanism similar to HTTPS Everywhere and will automatically router your URL request to the HTTPS version of the website if it is available.

- Automate your browser's attempt to connect to sites that use HTTPS. Firefox and Brave are good choices to configure such automation. Both have *HTTPS Everywhere*[7] built into them. If you cannot use either of these, you may be able to add HTTPS Everywhere as an extension to your preferred browser.

- Encrypt your entire online session using VPN. More on this on the *Internet Activity* chapter.

---

[7] *https://www.eff.org/https-everywhere/faq*

12.2 Choose A Browser

There are many web browsers on the market, with each placing different emphases on various features. The most popular browser for iOS is Safari. As this course is all about security, we focus on Safari, Brave, and Tor which are my favorites because of the security features they offer currently.

Why might you want to replace your current browser with another? There may be features–including additional security and privacy–offered by an alternative browser. Check out the features of browsers listed below:

Browser/Base	Platform	Price	Notable Features	Privacy
Brave Chromium	Android, iOS, macOS, Linux, Windows	Free	Built-in HTTPS Everywhere, Malware protection, Ad block	High
Chrome/ Chromium	Android, Chromebook, iOS, Linux, macOS, Windows	Free	We no longer advise using Chrome Browser due to growing security and privacy issues	Poor
Edge Chromium	Windows 11	Free (included with Windows 11)	The new Internet Explorer 11 is still IE, and Edge is not provided for Mac	Poor
Firefox Firefox (Gecko)	Android, iOS, Linux, macOS, Windows	Free (Open Source)	Add-ons, Privacy, History and Bookmarks can be shared between your devices running Firefox	High
Opera Chromium	Android iOS, Linux, macOS, Windows	Free	Built in VPN, sends info between devices, ad blocker	High

Browser/Base	Platform	Price	Notable Features	Privacy
Safari **Webkit**	macOS, iOS	Free	Built into Apple ecosystem, synchronized Bookmarks and History, forced HTTPS, Fingerprint blocking, malware protection	High
Tor **The Tor Project** **Firefox (Gecko)**	macOS, Linux, Windows	Free	Anonymous browsing, Built-in HTTPS everywhere, can access the dark web	Very High

12.2.1 Assignment: Secure Browsing with Brave

Brave was designed to be the most secure chromium browser available. It includes anti-malware, ad-blocking, HTTPS Everywhere.

In this assignment, you install the Brave browser to help ensure a more secure browsing experience.

Download Brave

1. Open the *App Store* app, search for then download *Brave Private Web Browser.*

Configure Brave

2. Open Brave.

3. Tap the ... (menu) icon > *Settings* > *Brave Shields & Privacy*. Configure to your taste. Shown below is my recommendation. When complete, tap *Back*.

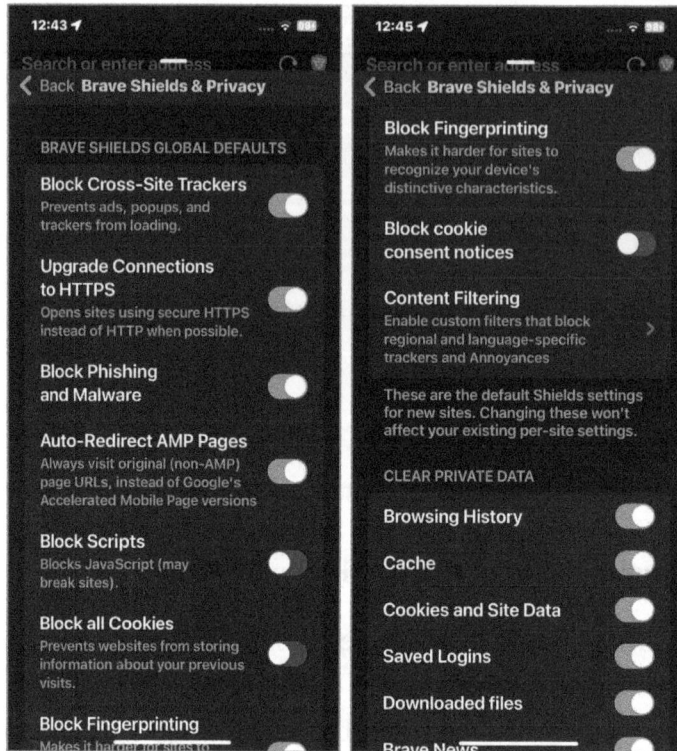

4. In the *Settings* screen, tap *Search Engines*. Configure to your taste. Shown below is my recommendation. When complete, tap *Settings > Back*.

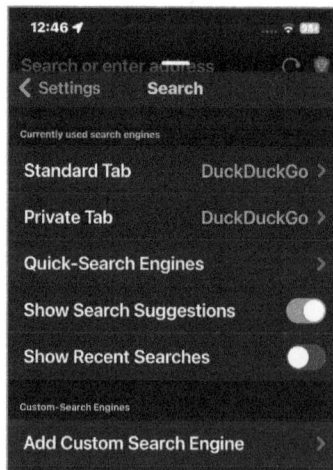

5. Open the *Settings.app > Brave*.

6. Configure to your taste. Shown below is my recommendation. When complete, exit out of *Setting.app*.

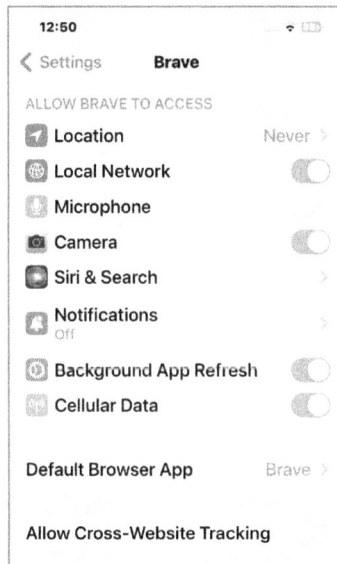

12.2.2 Secure Browsing with Safari

In this assignment, you harden Safari security.

1. Open *Settings*.app > *Safari*.

2. Configure to your taste. Shown below is my recommendation.

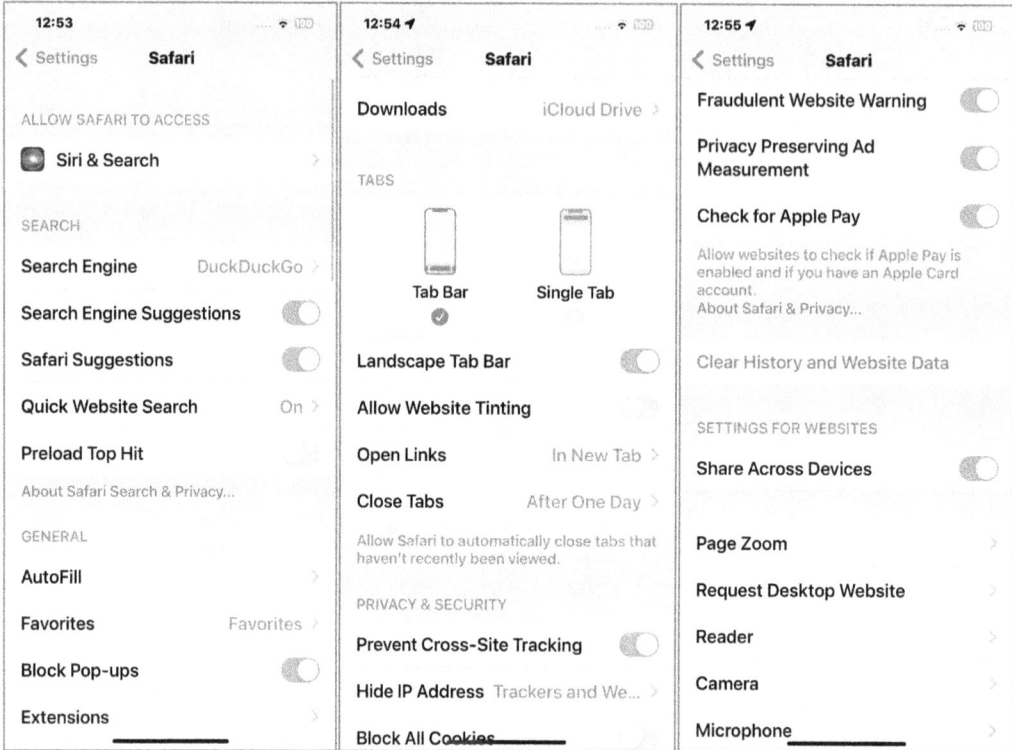

12:53		12:54		12:55	
‹ Settings **Safari**		‹ Settings **Safari**		‹ Settings **Safari**	

12:53 — ‹ Settings Safari

ALLOW SAFARI TO ACCESS

Siri & Search ›

SEARCH

Search Engine — DuckDuckGo ›

Search Engine Suggestions ⬤

Safari Suggestions ⬤

Quick Website Search — On ›

Preload Top Hit

About Safari Search & Privacy...

GENERAL

AutoFill ›

Favorites — Favorites ›

Block Pop-ups ⬤

Extensions ›

12:54 — ‹ Settings Safari

Downloads — iCloud Drive ›

TABS

Tab Bar ✓   Single Tab

Landscape Tab Bar ⬤

Allow Website Tinting

Open Links — In New Tab ›

Close Tabs — After One Day ›

Allow Safari to automatically close tabs that haven't recently been viewed.

PRIVACY & SECURITY

Prevent Cross-Site Tracking ⬤

Hide IP Address — Trackers and We... ›

Block All Cookies

12:55 — ‹ Settings Safari

Fraudulent Website Warning ⬤

Privacy Preserving Ad Measurement ⬤

Check for Apple Pay ⬤

Allow websites to check if Apple Pay is enabled and if you have an Apple Card account.
About Safari & Privacy...

Clear History and Website Data

SETTINGS FOR WEBSITES

Share Across Devices ⬤

Page Zoom ›

Request Desktop Website ›

Reader ›

Camera ›

Microphone ›

3. Tap *Siri & Search.* Configure to your taste. Shown below is my recommendation. When complete tap < *Safari.*

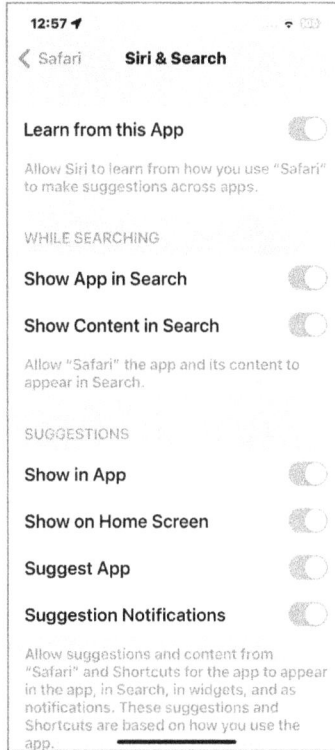

4. Scroll down to tap *Camera*, set to *Ask*, then tap < *Safari*.

5. Repeat previous step for *Microphone* and *Location*. When complete, tap < *Safari*.

6. Scroll down to tap *Advanced*. Configure to your taste. Shown below is my recommendation. When complete tap < *Safari* then exit *Settings.app*.

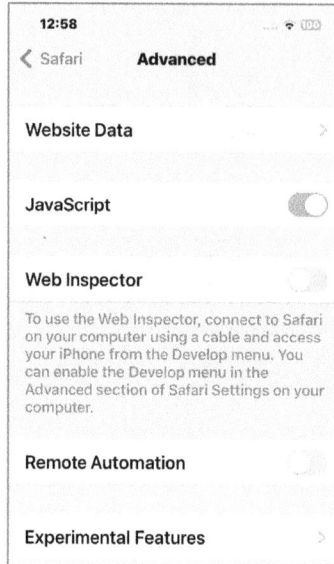

12.3   Private Browsing

Private Mode (Safari) and *Private Window* (Brave) are features that prevent any normally cached data from being written to storage while using a browser. This data includes browsing history, passwords, usernames, list of downloads, cookies, and cached files. Private browsing is an essential tool if you work on a computer where your account is shared (what's with that?), or if there is the possibility that someone else may examine your browsing habits.

- Note: This does not prevent your company IT department or Internet Provider from seeing or recording your browsing habits. To accomplish this, you must be using VPN (Virtual Private Network) on your device. More on that later.

12.3.1   Assignment: Brave Private Window

In this assignment, you enable Brave Private Window browsing.

1. Open Brave.

If you wish all your new pages to be opened in Private Window mode:

2. Tap *3 dot* menu > *Settings* > *Brave Shields & Privacy* > scroll down to enable *Private Browsing Only*. Then tap *Settings* > *Done*.

If you wish to choose if new pages are opened in Private Window mode or not:

3. Tap *3 dot* menu > *Settings* > *Brave Shields & Privacy* > scroll down to disable *Private Browsing Only*. Then tap *Settings* > *Done*.

4. Create a new Private Window by tapping the *Page* icon at bottom of screen > tap *Private* or tap +.

12.3.2   Assignment: Safari Private Browsing

In this assignment, you enable Private Browsing within Safari.

1. Open Safari.

2. Tap the *New Window* icon at bottom right of the screen > *Private*

12.4   Secure Web Searches

When you perform a search, the search criteria and sites visited may be collected and stored by the browser developer, search engine, your Internet Service Provider, and the trackers associated with sites visited. The cookies assigned from one website can communicate with other sites and webpages you open, allowing both the sites and advertisers to maintain logs of your activities and building a profile of your search history so that your search results are unique and tailored to your interests. Thus, if you query "gun," you can get ads related to this topic and beyond, such as militia websites. If you query baby animals, you can see postings more related to that, such as videos of kittens saved by dogs. The point is, not only are your habits and interests noted, but they are perpetuated by algorithms designed to send you to political and marketing bots from unknown parties. These sources can give different people different information, both true and false, and thereby manipulate our knowledge bases. It is no wonder that cultural division can result.

Not so with the *DuckDuckGo* search engine. DuckDuckGo's policy is to keep no information on user searches. Subsequently, your search activity is private, and all search results are identical for everyone. All users see the same data when searching on the same terms.

Browsers offer the option to make DuckDuckGo your default search engine. This is a big step towards providing a better level of privacy on the Web.

12.4.1 Assignment: DuckDuckGo for Brave Search

In this assignment, you change the default Brave search engine to the secure DuckDuckGo.

1. Open Brave.

2. Tap *3 dot* menu > *Settings* > *Search Engines*.

3. Set the *Standard Tab* to *DuckDuckGo*.

4. Set the *Private Tab* to *DuckDuckGo*.

5. Tap *Settings* > *Done* to exit.

12.4.2 Assignment: DuckDuckGo for Safari Search

In this assignment, you change the default Safari search engine to the secure DuckDuckGo.

1. Open *Settings.app* > *Safari*.

2. Set *Search Engine* to *DuckDuckGo*.

3. Exit *Settings*.

12.5 Clear History

By default, every browser maintains a full history of every site you have visited (when you are not in *Private* mode). Should someone gain access to your device, they can view your browsing history.

You can erase your entire browsing history in one tap. There is no recovery from clearing the browser history.

12.5.1 Assignment: Clear the Brave History

In this assignment, you clear the Brave browsing history.

- Warning: there is no recovery from this action. If you wish to keep your history, pass on this assignment.

1. Open Brave.

2. Tap *3 dot* menu > *Settings* > *Brave Shields & Privacy.*

3. Scroll down to *Clear Private Data* area, select the data to be cleared, then tap *Clear Data Now.*

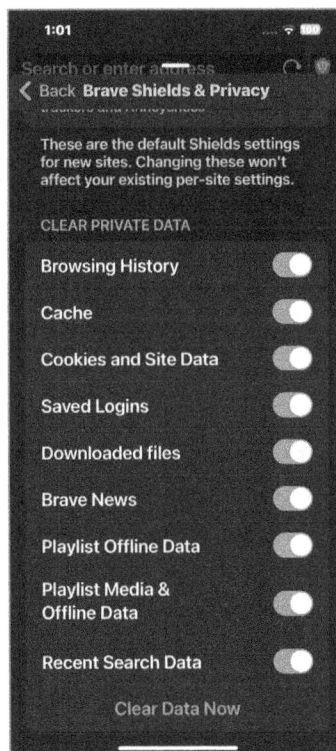

12.5.2   Assignment: Clear Safari History

In this assignment, you clear the Safari browsing history.

- Warning: there is no recovery from this action. If you wish to keep your history, pass on this assignment.

1. Open *Settings.app > Safari.*
2. Scroll down to tap *Clear History and Website Data.*
3. At the warning dialog, tap *Clear History and Data.*
4. Exit *Settings.*

12.6   Browser Extensions

One of the great advances in personal computer software development was the concept of plug-ins or extensions[8]. These small strings of code add functionality to the host application. In the case of web browsers, a function may be anything from the ability to encrypt web-based email to the ability to view proprietary video formats.

The bad news about extensions is that they run with the full power of the host application. This means that a malicious extension may have the power to secretly redirect your web browser to fake websites (such as a phony copy of your bank website), or harvest all your passwords, monitor your purchases, etc.

There are many malicious extensions. It is vital to install only those that you need to install, to know which are installed, and to rid yourself of unnecessary extensions.

To help ensure internet security and personal privacy, Apple has disallowed browsers running under iOS from installing plug-ins or extensions.

---

[8] *https://en.wikipedia.org/wiki/Plug-in_(computing)*

12.6.1 [ChromeOS, macOS, and Windows] Assignment: Find, Remove, and Add Brave Extensions

12.6.2 [macOS] Assignment: Find, Remove, and Add Safari Extensions

12.7 Fraudulent Websites

As of this writing, there are almost 2,000,000,000 active websites[9]. Within that, there may be millions (hundreds of millions?) of fraudulent websites. Of the diverse types of fraud found on the Internet[10], among the most common are websites that misrepresent who they are. This may be in the form of appearing like Bank of America, but with a URL of perhaps https://bankofamerica.cm, instead of the true https://bankofamerica.com. In this case, the *https* means the site is encrypted, not necessarily a valid or safe site. A criminal is hoping for someone to make accidentally type `.cm` instead of `.com`. Once at the fake site, you would enter your account and password as usual. The difference is that this time, a criminal now obtains your credentials–and all your money within minutes.

As a side note, in this specific example at the time of this writing (for a much earlier version of MacOS), this URL *is* a scam site. But not for the scheme mentioned. When I went to *http://bankofamerica.cm*, I was routed to the following:

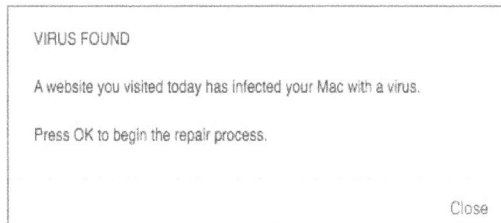

VIRUS FOUND

A website you visited today has infected your Mac with a virus.

Press OK to begin the repair process.

Close

---

[9] *https://www.internetlivestats.com/total-number-of-websites/*
[10] *https://en.wikipedia.org/wiki/Internet_fraud*

If we look at the full URL, it is: http://apple.com-----systemmessenger.com/dgkg/?city=Albuquerque&region=New%20Mexico&country=US&ip=71.222.135.33&isp=Qwest%20Communications%20Company%20Ll c&os=OS%20X&osv=OS%20X%2010.11%20El%20Capitan&browser=Safari& browserversion=Safari%209&voluumdata=BASE64dmlkLi4wMDAwMDAwNi0 1Yzg0LTRjNjYtODAwMC0wMDAwMDAwMDAwMDBfX3ZwaWQuLmRl...

From this URL, we can see that the criminal site attempts to appear as though it is Apple reporting that I have a virus. They have also discerned my city, state, IP address, Internet provider (Qwest Communications), and that I am using OS X 10.11 El Capitan with Safari version 9.

If I were the typical user, I'd think there was a virus present and press the *OK* button as recommended. You also may have noticed the criminal was bright enough to do all of this, but not bright enough to put an *OK* button in the script!

So, I press the *Close* button. I'm presented with a new window:

Hey, who needs an *OK* button when the *Close* button provides the intended scam! So, let's see what happens when I tap the *Scan Now* button:

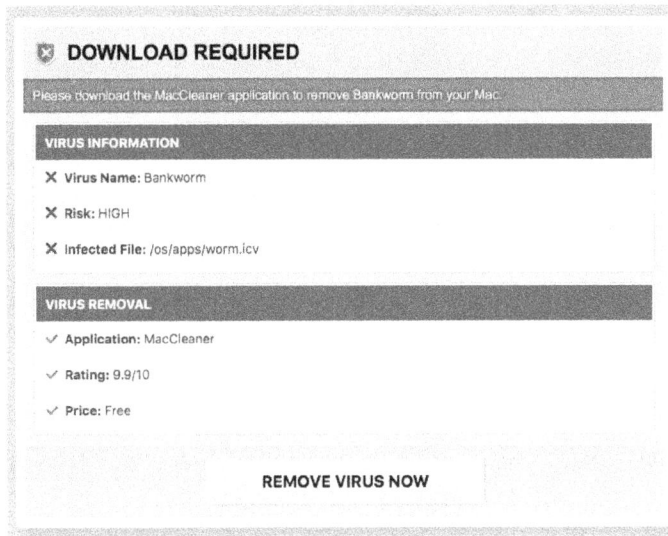

It appears the scammer thinks I am infected with the *Bankworm* virus, and the infected file is in /os/apps/worm.icv. The only real problem I see is that there is no such file, and no such directory.

But they are offering a free solution to my non-existent problem. Let's see where that takes us by tapping the *Remove Virus Now* button:

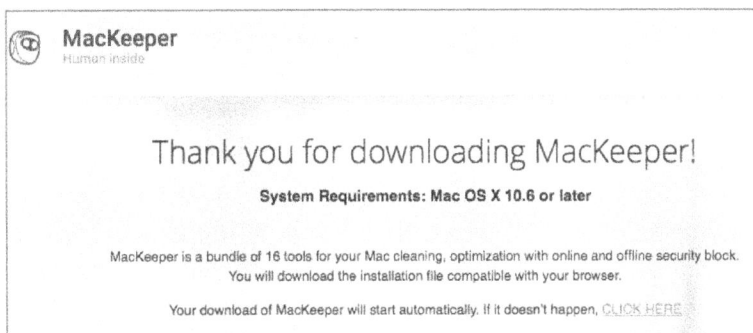

MacKeeper? Really! This product lost a class action lawsuit for deceptively advertising its functionality[11].

If you have followed along so far, just trash the MacKeeper download.

---

[11] *https://topclassactions.com/lawsuit-settlements/closed-settlements/94767-mackeeper-class-action-settlement/*

So, how to protect yourself against fraudulent sites? We go through the few steps that can be taken, but the most important tool is your awareness.

12.8 Do Not Track and Fingerprinting

Most websites track which pages you visit, how long you stay on each page, and other metrics to better understand their visitors. That is a little creepy. Imagine going to the library and having a librarian look over your shoulder as you scan the card catalogue and record each of the books and pages you glance at.

Now let's take the analogy further. You leave the library and go across town to have lunch. You look around and the same librarian still is watching and recording everything you looked at on the menu and what you ate.

Later you go on a date, and the librarian is sitting right behind you in the theater, noting who you are with, what scenes you react to, and more.

Web browsing isn't much different–except the snoop is normally invisible in the form of cookies, trackers, and browser fingerprinting.

Any website can initiate cookies on your browser. These keep a record of the pages you visit on the site. But they have evolved to report all the other places you visit and things that you do. Therefore, you can visit Amazon, look up my books, quit the web browser, launch it, go to Facebook, and see an ad for my books!

In addition, there are third-party trackers that are integrated with many sites. When you visit one site, you are uniquely identified. In this way, when you visit another site, the tracker knows who you are. Trackers can piece together a solid psychological, sociological, and financial profile on you.

There is the option to disable cookies, but most of your websites demand cookies be enabled to visit the site. And there are many ways to track you, even without cookies. To get a good sense of the philosophy of the trackers, look at this slightly dated *Guide to Cookieless Tracking*.[12]

---

[12] *https://www.ionos.com/digitalguide/online-marketing/web-analytics/browser-fingerprints-tracking-without-cookies/*

Trackers are invisible by default. However, we do have tools to thwart them. Some web browsers have a preference setting to ask web sites not to track you. Notice the term is *ask*. Good luck with that. Of course, you can turn off cookies but often that means the site will refuse to deal with you. You must decide how much you need the product or information offered by the site.

Device fingerprinting[13] is a bit more of a problem. When opening a webpage, it is possible for the site (or a search engine) to uniquely identify your device amongst the hundreds of millions of other devices on the internet. Once you have been fingerprinted, it is easy to track your activity across the internet. As ghastly as this is, there is a fix to make your fingerprint appear more generic.

There are promises from browser developers to build in fingerprinting blocking technology. But it is currently only reasonably implemented in Safari, Tor, and Brave. There are browser extensions to block fingerprinting as well.

Safari calls their fingerprinting blocking *Intelligent Tracking Prevention*. It includes the following features:

- Blocks third-party cookies, preventing advertisers and websites from following you around the internet through cross-site tracking.

- Deletion of third-party cookies and website data unless you have directly interacted with the content provider.

- Blocks web tracking through machine learning.

- Provides a substitute IP address, hiding your actual IP address.

The only 100% solution at this time is to use a bootable thumb drive with Tails[14] installed, then do nothing to alter the drive or OS setup. When booting from Tails, you look exactly like all other Tails users, and it is impossible to fingerprint you.

12.8.1 Assignment: View Your Device Fingerprint

---

[13] *https://en.wikipedia.org/wiki/Device_fingerprint*
[14] *https://tails.boum.org/*

Each device on the internet is unique based on its combination of features, traits, fonts, settings, etc. This can be used to identify the device, then to track it across the internet. This is called a *device fingerprint*[15].

In this assignment, you view your device fingerprint.

1. Open a browser, then go to *https://amiunique.org*.

2. Scroll down to select *View My Browser Fingerprint*. In a minute or two a summary of your unique fingerprint displays.

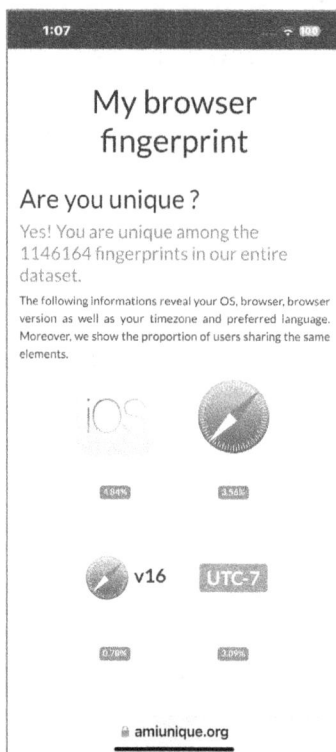

3. Scroll down to see all your fingerprint attributes.

4. Open a new browser window, then go to *https://panoptitap.eff.org*. This site is similar in function to *amiunique.org* and provides some unique information.

5. Enable the *Test with a real tracking company* checkbox.

---

[15] *https://en.wikipedia.org/wiki/Device_fingerprint*

6. Select the *Test Your Browse* button. After a few minutes, a results summary appears.

7. Scroll down to the *Detailed Results* area > *Select a characteristic* pop-up menu, then tap *Detailed View* to display to view all your fingerprint details.

12.9 Web Scams

Over the past couple of years, a new type of scam has become popular. Instead of directly compromising the user computer, web sites are compromised or deliberately designed to be malicious.

When you visit such a site, you may receive a pop-up window stating something to like: Your computer has been found to be infected with XX viruses. Please call Apple at XXX-XXX-XXXX to have this infection removed.

Upon calling the provided toll-free phone number (which, of course, is not Apple, but the scammer), with your permission, they install remote control software. After looking around your computer, they assure you they can remove the malware for only $$$. Just recite to them your credit card information.

There are two big problems here. First, a criminal installed remote control software that allows the criminal access any time they wish. They can access your usernames, passwords, banking, and your other information. The second is that they now have your credit card information.

Do you think you are too smart, educated, or street-smart to fall victim? These attributes may put you at *greater* risk![16, 17]

12.9.1 Assignment: Recovering from a Web Scam

In this assignment, you examine your browsers for possible modifications. What to do if a scam happens to you?

---

[16] *https://www.marketplace.org/2011/09/30/why-even-smart-people-fall-scams/*
[17] *https://assets.aarp.org/rgcenter/econ/fraud-victims-11.pdf*

1. Do not call the number offered by the scammer! These are criminals.

2. In most cases, the malicious website has modified your web browser preferences to make the malicious page your home page.

   Safari

   iOS Safari has no setting for home page, it opens to the last page you were on. You can clear this last page by:

   a. Open Safari.

   b. Tap the *Pages* icon in the bottom right corner.

   c. Tap the *x* icon to the far left of the page URL.

   Brave

   iOS Brave has no setting for home page, it opens to the last page you were on. You can clear this last page by:

   a. Open Brave.

   b. Tap the *Pages* icon near the bottom right.

   c. Tap the *x* icon in the top right corner of the page thumbnail.

3. Quit your browser.

4. Open browser to test. You should no longer have the malicious page open.

5. Repeat steps 1-4 for each of your browsers.

Check for DNS changes

Some malicious sites change your DNS settings. In doing so, they can redirect your browsing to fake sites they control but look like the real deal. Their hope is you enter sensitive data they can then use (passwords, bank account numbers, etc.)

6. Open *Settings > Wi-Fi*.

7. Tap the *Info* icon for your current Wi-Fi network.

8. Scroll down to *Configure DNS*. If it is set to *Automatic,* you are OK. If it is set to *Manual,* tap *Manual*, then inspect the IP addresses of your DNS servers. If they are not what you had previously configured, reset them.

- Note: By default, network devices take on the DNS server addresses fed to them by the network router. In turn, by default, routers take on the DNS server addresses fed to them from their Internet provider. I recommend using the Cloudflare DNS server, with addresses of 1.1.1.1 and 1.0.0.1. These servers do not track your activities and are currently the fastest available.

9. Select *Back*.

10. Exit *Settings*.

Block Malicious Sites

Brave

11. Open Brave.

12. Select *3 dot* menu > *Settings* > *Brave Shields & Privacy*.

13. Scroll down to enable *Block dangerous sites*.

14. Exit *Settings*.

Safari

15. Open *Settings.app* > *Safari*.

16. Scroll down to enable *Fraudulent Website Warning*.

17. Exit *Settings*.

18. Done!

12.10 Tor

- Note: As of this writing, Tor has not yet been migrated to iOS. Although there are many "Tor" apps on the App Store, none have been authorized by the Tor Project[18] team, and most should be considered suspect. The Tor Project only recommends the iOS *Onion Browser*

---

[18] *https://www.torproject.org/*

Tor[19] is a technology developed by the US Department of the Navy that enables anonymous web browsing. Anonymous browsing is accomplished by first encrypting your data destined for the internet, then routing it through multiple *onion* routers that make up the *onion network*[20]. Each router only knows which router the data packets came from, and which router to forward to, so as your data exits the final router it is not possible to trace the packets back to their origins.

Anonymous browsing has long since been released to the open-source community for the public to use in the form of the *Tor Browser*. The Tor browser is a stripped-down browser with the unique ability to communicate with the onion network.

Many people in the security community are strong supporters of Tor, including Edward Snowden. Entire books have been written on just Tor. I'm not so sadistic as to subject you to that. However, keep in mind that although Tor may keep your Internet activities anonymous to criminals, the US government (and likely other governments) can compromise Tor.

Tor advantages:

- Strong anonymity for all activity on the Internet.

- Can be used with Tails[21] which is a bootable, self-contained, flash drive. Tails can run on most Windows, Linux, and Apple (Intel Macs only) computers and it leaves behind no trace of activity.

- The bootable Tails flash drive can be immediately disconnected from the host computer, causing the computer to erase memory of all trace of your session, and reboot.

Tor disadvantages:

- Tors was developed by the US Department of the Navy. It is possible there are back doors only the government knows about.

- The US government has been forthright about having its own Tor relays in place, which enable it to monitor online activity. Not a big deal if you only wish to be anonymous to criminals. It is a big deal if you wish to be

---

[19] *https://en.wikipedia.org/wiki/Tor_(anonymity_network)*
[20] *https://en.wikipedia.org/wiki/Onion_routing*
[21] *https://tails.boum.org*

anonymous while performing black-market deals for Aunt Rose's raisin Noodle Koogle recipe.

- The onion network and Tor help ensure your anonymity by bouncing your web traffic around at least three random onion routers. This can dramatically reduce your browsing performance.

These features make Tor ideal for those in oppressed countries, journalists working undercover, and anyone who may need to use someone else's computer and leave no trace behind.

Tor works by encrypting your packets as they leave your computer, routing the packets to a Tor relay computer hosted by thousands of volunteers on their own systems, many of which are co-located at ISPs. The relay knows where the packet came from, and the next relay the packet is handed to, but that is all. The user computer automatically configures encrypted connections through the relays. Packets pass through several relays before being delivered to the intended destination. Tor uses the same relays for around 10 minutes, then different relays are randomly selected to create the next path for 10 minutes.

Alas, there is no free lunch. The encryption process and the relay process combine to create *latency*, which means a delay in processing. Most users experience around a four-fold performance degradation. So, if accessing a web page without Tor normally takes 3 seconds, it may take 12 seconds with Tor.

Even though Tor does as good a job as anything to keep you anonymous on the Internet, you must take precautions to protect your identity. These steps include:

- Do not enable JavaScript when using Tor. JavaScript has been used to track users within the Tor network.

- Do not reveal your name or other personal information in web forms.

- Do not customize the Tails boot flash drive. This creates a unique digital fingerprint that can be used to identify you.

- Connect to sites that use HTTPS, so your communications are encrypted point to point.

Tor by itself is at best a partial solution. It can protect your anonymity while surfing the web. At the very least, this still leaves email (if you are not using web-based email) and messaging to be secured. A bigger issue is what to do when you need to use a computer and leave no trace behind on that system. This is where Tails comes into play.

Tails is a Linux Debian fork designed with two primary purposes:

- Provide a highly secure operating system in a format that can be booted from either DVD or thumb drive on almost any PC or Apple computer, and

- Include the tools and applications necessary to provide a secure, anonymous Internet experience.

What this means is that, with Tails, you can create a thumb drive that has an operating system capable of booting almost any computer, whereby you can run Tor for secure anonymous Internet activity, send and receive email that is securely encrypted with GPG/PGP, and message with others in complete privacy. Then, when you remove the Tails thumb drive, there is absolutely no record of your activity on either the computer *or* the thumb drive.

For those of you chomping at the bit to just use Tor, we start there. When your curiosity has been satisfied, please take the next step to learn Tails[22].

12.10.1 Assignment: Use Tor for Anonymous Internet Browsing

Tor is a stripped down, simplified web browser, providing the highest level of privacy while browsing. The trade-off is lack of flexibility, as modifying Tor's preferences or adding extensions makes your identity unique.

The *Onion Browser* is a free Tor Browser for iOS, allowing you to browse the web with a reasonable level of anonymity. Although not officially supported by the Tor Project, the Tor Project only recommends this browser for iOS.

In this assignment, you install Onion Browser on your iOS device.

Record Your Public IP Address

1. As a first step, we need to know our public IP address. This information is used a few steps away to verify Tor has hidden our address. Open a web browser to *https://whatismyip.com*. Write down *Your IP*.

---

[22] *https://tails.boum.org*

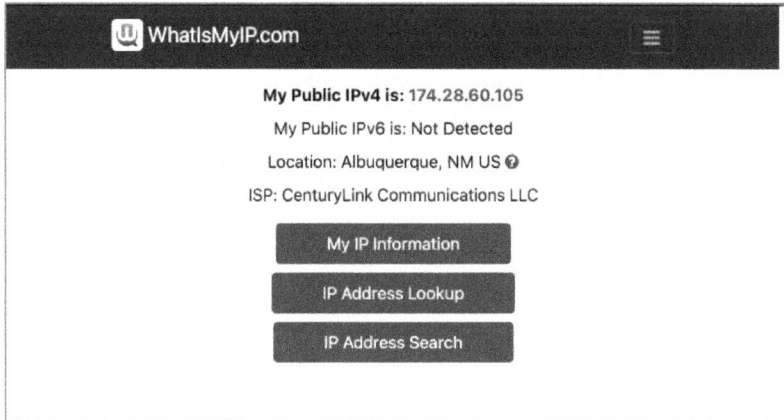

Install Onion Browser

2. Open *App Store*.app, search for then download *Onion Browser*.

3. Once downloaded, open Onion Browser.

4. At the *Connect to Tor for private browsing* screen, select either *Connect to Tor*, or *Configure Bridges*. In almost all cases you should select *Connect to Tor*.

5. At the *Define Default Security Level* screen, select how you wish to balance security versus performance. Then tap *Start Browsing*.

6. The *Onion Browser* home page appears.

7. In Onion Browser, return to *https://whatismyip.com*. Notice that you now have a different IP address. To the internet, you appear to be on an entirely different network.

12.10.2 Configure Onion Browser Preferences

One of the first things one should do when launching an application for the first time is to configure its preferences. No different for Onion.

In this assignment, you configure Onion preferences.

*   Prerequisite: Completion of the previous assignment, or having Onion installed on your device.

1. Open Onion, then select the *Gear* icon > *Preferences* > *General.* Configure to your taste. My settings are below:

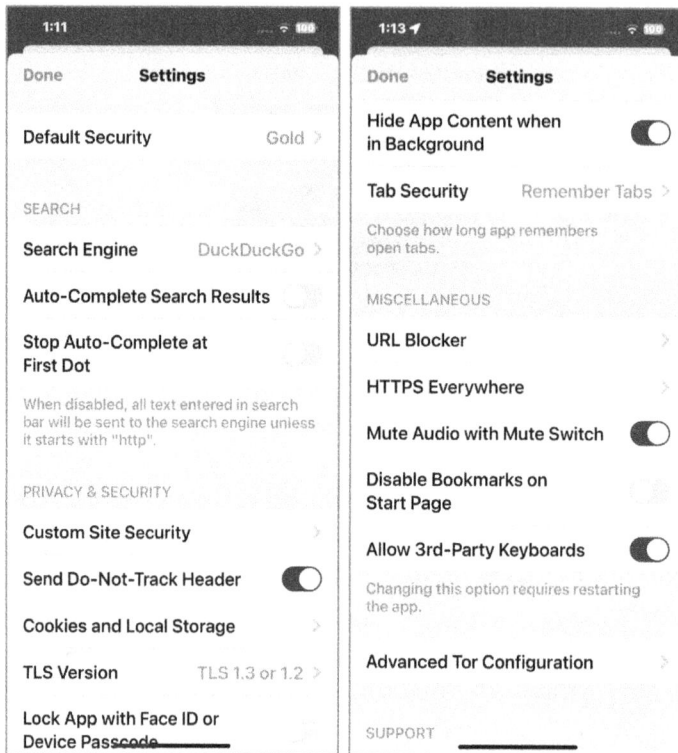

2. When complete, tap *Done.*

You are now connected to the Onion network. You can use Onion Browser just as you would any non-Tor browser to browse the internet.

When using Tor and the onion network, there are other steps that one must take to ensure privacy is maintained. These include:

*   Use the Tor Browser. If you are concerned about protecting your privacy and security, do not use other browsers.

*   Do not torrent over Tor. If you wish to file-share via torrent, do not use Tor. Torrent is painfully slow, it slows down others using the Tor network, and in many cases, torrent software bypasses all the security and anonymity precautions built into Tor and is broadcast unencrypted.

- Do not enable or install browser plugins in Tor. Tor is designed to protect your security and anonymity. Many innocuous-looking plugins break that security.

- Use HTTPS versions of websites. Tor has *HTTPS Everywhere* built in (more on HTTPS Everywhere later in this book). It forces a secure connection if a website has an option for https. This enables a point-to-point encryption between your computer and the web server.

- Do not open documents downloaded through Tor while online. Many documents–particularly .doc, .xls, .ppt, and .pdf–contain links, macros, or resources that force a download when the document is opened. If they are opened while Tor is open, they reveal your true IP address, and you lose your anonymity and security. If you are concerned about these issues, we strongly recommend doing this instead:

  o Internet. This prevents any malicious files from "phoning home" or infecting your computer.

  o Install a Virtual Machine (VM) such as Parallels, Fusion, or VirtualBox, configured with no network connection, and open documents within the VM. This is an alternate way to prevent malicious files from phoning home or infecting your computer.

  o Or use Tor while within Tails. This is an alternative way to prevent malicious files from phoning home or infecting your computer.

- Use bridges and/or find company. Tor cannot prevent someone from looking at your Internet traffic to discover you are using Tor, even if they cannot determine what it is you are doing. If this is a concern for you, reduce the risk by configuring Tor to use a *Tor Bridge relay* instead of a direct connection to the Tor network. Another option is to have many other users running Tor on the same network. In this way, your use of Tor is hidden.

Wahoo! You are now on Tor, completely anonymous and encrypted on the Internet.

12.10.3 Assignment: Browse with Onion

In this assignment, you browse a site with both its www and onion address.

- Prerequisite: Installation of Onion.

1. Open Onion browser.

2. In the address field, enter `https://torproject.org`. This is the home for the Tor developers. The site opens just as it would with any other browser.

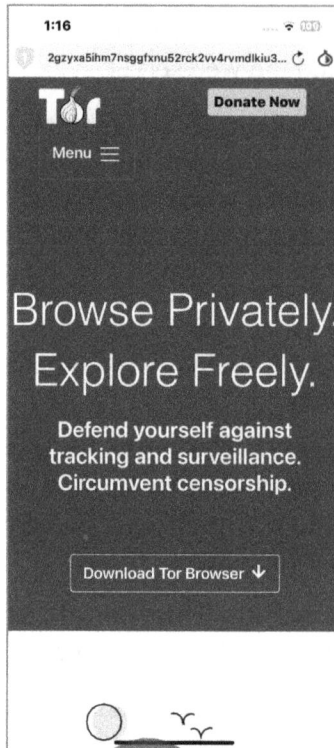

3. Note the address shown in the address field. In this case, the browser automatically redirected from the surface web site to the onion dark web site. There are only a few sites for which this is done. You typically need to know the onion address of a site.

4. Quit Onion.

12.10.4 [macOS and Windows] Brave Private Window with Tor

12.11 Surface Web, Deep Web, and Dark Web

The internet can be viewed as having three layers: *Surface, Deep, and Dark.*

The *Surface Web* is anything that can be indexed by a typical search engine like Google. A good way to visualize this is to go to a web site–for this example, Amazon.com–then tap any links found on the site. This is how a search engine works. It follows links.

The *Deep Web*[23] is anything that cannot be indexed by a typical search engine like Google. A good way to visualize this is to go back to Amazon.com, then find the price of a specific product from 15 years ago–but you are limited to doing so by only using links. It cannot be done. Such information could be found using the search feature within Amazon, but that isn't how a search engine works. It can only use links. This product information would be considered Deep Web content.

The *Dark Web*[24] is that portion of the Internet content that requires specific software, configurations, or authorization to access. The Dark Web is a small part of the Deep Web. The most common tools used to access the Dark Web include Tor[25], I2P[26], and Freenet[27].

Tor not only allows you to have anonymous access to your regular web sites, but it is also one of the few tools offering a gateway to the Dark Web. Web sites on the dark web are also called *Onion sites,* as they end with *.onion.*

Although the dark web is primarily thought of as a collection of sites to sell illegal products and services, there are also good and responsible uses for it. For example, in repressive countries such sites provide an avenue for freedom workers and reporters to securely exchange information with sources (Ed Snowden did this), and there are sites to provide resources for whistleblowers.

As the dark web is not indexed by Google, Bing, or any other standard search engine, how do you go about discovering its resources? The list is in constant

---

[23] *https://en.wikipedia.org/wiki/Deep_web*
[24] *https://en.wikipedia.org/wiki/Dark_web*
[25] *https://en.wikipedia.org/wiki/Tor_(anonymity_network)*
[26] *https://en.wikipedia.org/wiki/I2P*
[27] *https://en.wikipedia.org/wiki/Freenet*

flux. One of the better ways of finding dark web directory listings is a web search of *dark web directory.*

12.12 Hacked Accounts

Do you know if you've been pwned?

"WHAT!?!" is probably the first thing that just went through your mind. No, it's not a typo. *Pwn*, as defined in the dictionary, is to be totally defeated or dominated. Although most used when trouncing your online game opponent, it is also used to describe when your email or online accounts have been hacked.

Unfortunately, there is a pretty good chance that you have been pwned!

Several websites track email and online account breaches. My favorite is *haveibeenpwned.com.*

12.12.1 Assignment: Find Your Hacked Accounts

In this assignment, you search the haveibeenpwned.com database to discover if any of your online accounts have been hacked/pwned.

1. Open a web browser to *https://haveibeenpwned.com.* The home page appears.

2. Enter your email address, then tap the *pwned?* button.

3. In a few seconds, the results display.

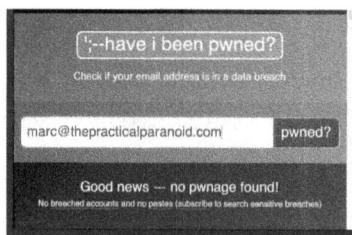

4. Make a note of the sites with breaches. In most cases, your concern should focus on breaches that include account name or email address, and password.

- Note: the site may report there have been *pastes*. A paste is information about your pwned account that has been "pasted" to a surface web site such as Pastebin.com. These are services favored by hackers.

5. Close your browser.

12.12.2 Assignment: What to Do After You Have Been Hacked

In this assignment, you take action to repair any found breaches.

- Prerequisite: Completion of the previous assignment.

1. If the breach is a specific website, open a web browser. Then go to the breach site.

2. Change your account password, following these best practices:

    - Passphrase is a minimum of 15 characters in an easy-to-enter phrase.

    - Use each password/passphrase for only one site. Should a site become compromised, and your password harvested, many automated hacking systems then use your credentials at every bank, online store, etc. they know of to see if you, like most folks, use one password for everything.

    - Keep a secure record of your passwords/passphrases. I use Bitwarden as my password manager. You can also use a current version of Excel to create an encrypted spreadsheet, which you must update whenever you change your access information.

    - Only enter a username and password when operating in a secure web page (https).

3. Repeat step 2 for each breached site.

4. If the breach is not of a specific website or service, there is nothing you can do. In most of these cases, massive collections of breaches from unknown sources are placed on the dark web for sale, or just given away. To repeat: it is so important to not use the same password for more than one site or service.

12.13 Ad Targeting Information

As you browse the Internet, many of the websites you visit track your activity. This information can then be knitted together with the tracking from all your online activity to create a valuable package about who you are, your likes, and your private life.

Although it may not be possible to eliminate all traces of this tracking activity, you do have the power to significantly reduce it.

12.13.1 Assignment: Apple Advertising

Apple provides some transparency into its use of ad targeting information when you visit Apple sites such as App Store, Apple News, and Stocks. For a full disclosure on Apple advertising and privacy policies, visit *https://support.apple.com/en-qa/HT205223*.

In this assignment, you reduce your ad targeting information available to Apple.

1. Open *Settings.app > Privacy > Apple Advertising > View Ad Targeting Information*.
2. When complete, return to *Settings.app*.

12.13.2 Assignment: Turn Personalized Ads Off

In this assignment, you turn off Apple's ability to delivery relevant ads to you.

- Note: This does not reduce the number of ads you receive.
1. Open *Settings.app > Privacy > Apple Advertising*.
2. Turn *Personalized Ads* to *Off*. When complete, return to *Settings.app*.

12.14 Web Browsing Lessons Learned

☐ Websites using http are not secure or private. All data between your device and the server is in clear text.

☐ Websites using https are secure and private. All data between your device and the server is encrypted. The *lock* icon denotes https.

☐ Browser choice is critical to security and privacy. Some browsers exist to harvest your browsing habits.

☐ Private browsing does not prevent your router, business, or internet provider from recording your browsing activities.

☐ Most search engines are funded by selling your browsing activity records.

☐ Clearing your browser history prevents knowledge of your browsing activities if someone has access to your account on your device.

☐ Browser extensions can monitor your browser activities.

☐ Your browsing activities can be monitored and followed from site to site, from browser to browser and even device to device, based on your device fingerprint.

☐ The internet can be viewed as having three layers. The surface web is anything that can be indexed by a typical search engine. The deep web is anything that cannot be indexed by a typical search engine. The dark web are services and servers accessible only using the secure onion network and onion network compatible browsers, such as Tor and Brave.

☐ It is recommended to review haveibeenpwned.com monthly in case your internet accounts have been compromised.

☐ You can reduce the amount of tracking done by Apple by configuring *Settings > Privacy > Apple Advertising*.

☐ Brave is one of the most secure and private browsers available.

☐ Tor may be the most secure and private browser. It uses both encryption and bouncing from router to router to secure your data and shield your identity.

☐ New in iOS 15 is that Safari will automatically connect to the HTTPS version of a site if available.

13 Email

Human beings the world over need freedom and security that they may be able to realize their full potential.

–Aung San Suu Kyi[1], Burmese opposition leader and chairperson of the National League for Democracy in Burma

What You Will Learn in This Chapter

- Avoid Phishing
- Use email encryption protocols
- Use TLS encryption
- Use HTTPS encryption
- Create end-to-end secure email with ProtonMail
- About PGP/GPG and S/MIME
- Create end-to-end secure email with Paubox
- Configure SPF, DKIM, and DMARC records
- Use email 2-Factor Authentication

What You Will Need in This Chapter

- [Optional] administrator access to your private domain control panel and DNS settings.
- [Optional] email account on private domain.
- [Optional] Paubox account.

---

[1] *https://en.wikipedia.org/wiki/Aung_San_Suu_Kyi*

- [Optional] possess a browser based Outlook.com email account.

13.1 The Killer App

It can be rightfully argued that email is the killer app that brought the Internet out of the geek world of university and military usage and into our homes. Most email users live in some foggy surreal world with the belief they have a God or constitutionally given right to privacy in their email communications.

No such right exists. Google, Yahoo!, Microsoft, Comcast, or whoever hosts your email service is likely to turn over all records of your email whenever a government agency asks for that data. In many cases, your email is sent and received in clear text so that anyone along the dozens of routers and servers between you and the other person can clearly read your messages. Add to this knowledge recent revelations about PRISM[2] (2007), also known as SIGAD US-984XN, where the government does not have to ask your provider for records, the government simply *has permission to access* your records.

Let's talk about what you can do to minimize your loss of privacy.

13.2 Phishing

Phishing[3] is epidemic on the Internet. Phishing is the attempt to acquire your sensitive information by appearing as a trustworthy source. This is most often attempted via email.

The way the process often works is that you receive an email from what appears to be a trustworthy source, such as your bank. The email provides some motivator to contact the source, along with what appears to be a legitimate link to the source website.

---

[2] *https://en.wikipedia.org/wiki/PRISM_(surveillance_program)*
[3] *https://en.wikipedia.org/wiki/Phishing*

When you tap the link, you are taken to what appears to be the trustworthy source (perhaps the website of your bank), where you are prompted to enter your username and password. If you do this, they have you. The site is a fraud, and you have just given the criminals credentials to access your bank account. In a few moments, your account may be emptied.

The key to preventing a successful phishing attack is to be aware of the *real* URL behind the link provided in the email.

The link that appears in an email may have nothing at all to do with where the link takes you. To see the *real* link, hover (do not tap) your cursor over the link. After 3 seconds, the *real* link pops up.

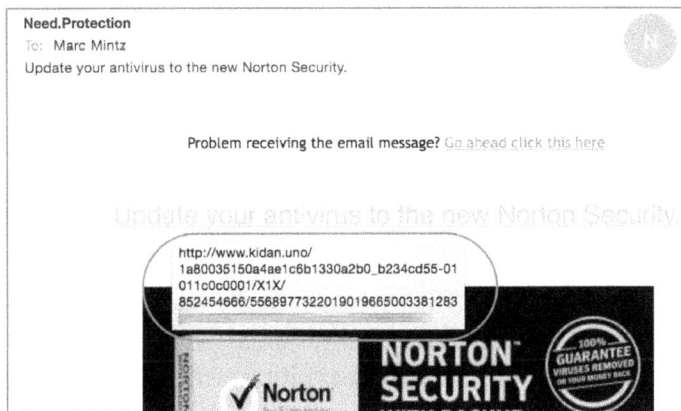

Some scams are getting more sophisticated in their choice of URL links and attempt to make them appear legitimate. For example, an email may say it is from *Bank of America*, and the link say *bankofamerica.com*, but the actual URL is *bankofamerica.tv*, or *bankofamerica.xyz.com*.

If you have any doubts at all, it is best to contact your bank, stockbroker, insurance agent, etc. directly by their known email or phone number.

13.3 Email Encryption Protocols

Three common protocols provide encryption of email between the sending or receiving computer and the SMTP (outgoing), IMAP (incoming), and POP (incoming) servers:

- **SSL**[4] (Secure Socket Layer)**,** now depreciated
- **TLS**[5] (Transport Layer Security), the successor to SSL
- **HTTPS**[6] (Hypertext Transport Layer Secure)

Understand that these protocols only encrypt the message as it travels between your computer and your email server, and back. Unless you are communicating with only yourself (sadly, as some programmers are prone), this does little good unless you know the other end of the communication also is using encrypted email. If not, then once your encrypted mail passes from your computer to your email server, it demotes to either the less secure SSL, or if the other end of the communications doesn't support that, demotes to clear text from your email server, through dozens of Internet routers, to the recipient email server, and finally onto the recipient's computer.

13.4 Transport Layer Security (TLS)

Although Secure Socket Layer (SSL) was originally considered secure, it has been broken and should no longer be used for email that is sensitive, secure, or related to healthcare, legal, government, or military sites. Transport Layer Security[7] (TLS) is the successor to SSL. To use TLS, the following criteria must be met:

- Your email provider offers a TLS. Many do not. If your provider does not offer this, *run*, don't walk, to another provider. If you are not sure which to select, I'm a fan of Google mail.

- You are using an email application as opposed to using a web browser to access your email.

- Your email application supports TLS.

---

[4] *https://en.wikipedia.org/wiki/Secure_Sockets_Layer*

[5] *https://en.wikipedia.org/wiki/Secure_Sockets_Layer*

[6] *https://en.wikipedia.org/wiki/Https*

[7] *https://en.wikipedia.org/wiki/Transport_Layer_Security*

- Your email provider has enabled and configured your email service to use TLS (they may *offer* TLS, but it may not be *enabled* by default).

- You have configured your email application to use TLS. Most email applications now do this automatically. Apple Mail.app has gone to the point they have removed the preference setting for both SSL and TLS.

- Lastly, although not a requirement for TLS, a requirement to stall off hacking your password is that your email provider should allow for strong passwords, and you have assigned a strong password to your email. As noted, I recommend 15 or more characters. Unfortunately, many providers still limit you to a maximum of 8-character passwords.

13.4.1 Assignment: Determine if Sender and Recipient Can Use TLS

Email automatically downgrades to the lowest common security protocol.

In this assignment, you discover if both your own email and that of a recipient can use TLS email encryption.

1. Open a web browser, then go to *https://checktls.com*.

2. Scroll down the home page to the *Check Your, or Any, Email System* section.

3. In the *Check How You Get Email (Receiver Test) FREE* field, enter your email address. Then select the *Check It* button.

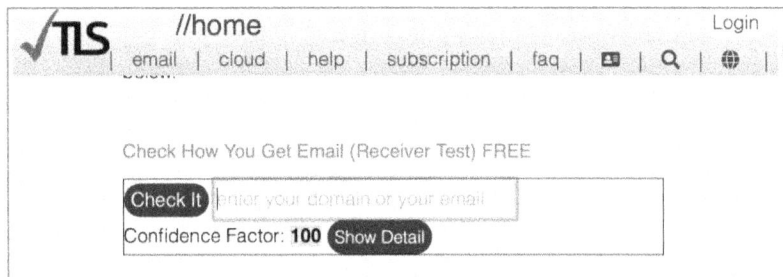

4. The website runs tests against the domain's mail servers (MX servers), then reports on their level of security. Tap the *Show Detail* button.

√TLS //email/test**To:** Visitor
| email | cloud | help | subscription | faq | 🖼 | Q | ⊕ |

CheckTLS Confidence Factor for "marc@thepracticalparanoid.com": 100

MX Server	Pref	Answer	Connect	HELO	TLS	Cert	Secure	From
aspmx.l.google.com [173.194.68.27:25]	1	OK (18ms)	OK (11ms)	OK (12ms)	OK (10ms)	OK (124ms)	OK (12ms)	OK (10ms)
alt1.aspmx.l.google.com [64.233.186.27:25]	5	OK (128ms)	OK (274ms)	OK (259ms)	OK (257ms)	OK (196ms)	OK (330ms)	OK (257ms)
alt2.aspmx.l.google.com [209.85.202.27:25]	5	OK (78ms)	OK (120ms)	OK (96ms)	OK (91ms)	OK (196ms)	OK (94ms)	OK (92ms)
aspmx2.googlemail.com [64.233.186.27:25]	10	OK (123ms)	OK (281ms)	OK (255ms)	OK (252ms)	OK (262ms)	OK (313ms)	OK (255ms)
aspmx3.googlemail.com [209.85.202.27:25]	10	OK (77ms)	OK (105ms)	OK (94ms)	OK (93ms)	OK (192ms)	OK (94ms)	OK (92ms)
Average		100%	100%	100%	100%	100%	100%	100%

5. If your *Test Results* are not 100% secure, either discuss this with your email provider for a resolution or change providers.

6. Repeat steps 1-4 using the domain of your recipient email address.

7. If the recipient's *Test Results* are not 100% secure, advise them to discuss this with their email provider, or change providers.

13.4.2 [Windows] Install Mozilla Thunderbird

13.4.3 [Windows] Configure Microsoft Outlook 365 for TLS

13.4.4 Assignment: Configure Mail.app to Use TLS

If you use a web browser for email, skip this assignment and move on to the next section where we configure your browser-based email to use HTTPS.

In this assignment, you verify if your email currently uses TLS.

1. Tap *Settings > Mail > Accounts.* All your configured email accounts display.

2. Tap the desired email account. This takes you to the account screen. Tap on the *Account <email address>* area.

3. Tap the *SMTP* field.

4. Tap the *Primary Server* entry field.

5. If *Use SSL* is on, good (Apple hasn't yet changed the interface to specify if the protocol is SSL or TLS). If it is off, do not activate it yet. You first need to see if your email provider supports TLS. If they do, then you need to ask what changes (if any) need to be made to your iOS email settings to use encryption, then make them (in most cases there is nothing to do but activate the *SSL* switch). If your provider doesn't support TLS, *run* to another provider.

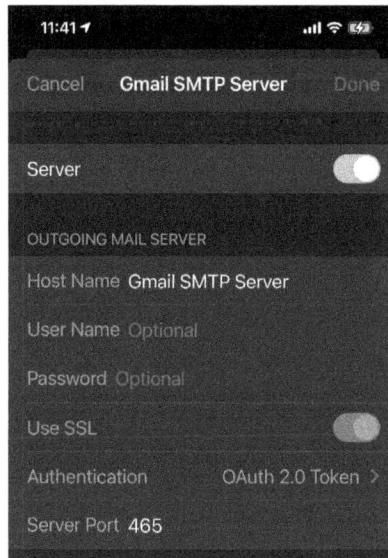

13.4.5 [Optional] Assignment: Configure Forced TLS With Paubox

One option to force bi-directional, HIPAA-compliant, secure, encrypted TLS email communications is by using Paubox[8]. Paubox is a third-party email server that can be used with your existing G-Suite, Office 365, and Microsoft Exchange email, as long as you have a private domain name associated with your email (i.e.: marc@thepracticalparanoid.com, not marcmintz@gmail.com).

---

[8] *https://paubox.com*

When sending email through Paubox, your interface doesn't change – you can still use the email client or web browser you have always used. What does change is that your email now goes through Paubox. Paubox servers check with the recipient email server. If the recipient server supports TLS, your email is sent to the recipient as usual.

If the recipient email server does not support TLS, the email is instead stored on the Paubox server. A message is sent to the recipient inbox stating that an email is waiting for them on the Paubox server, and all they need do is tap the button to view the email.

The recipient can reply to the email, which is then sent end-to-end encrypted with TLS. But understand that if the recipient simply emails to you, and their server doesn't support TLS, the email is sent without end-to-end TLS encryption.

In this assignment, you create a free 14-day trial Paubox account.

- Prerequisite: An email with a private domain name, not a generic such as @gmail.com.

- Prerequisite: Access to the domain name DNS records control panel.

Create a Paubox Account

1. Open a browser to *https://paubox.com*.

2. Tap the *Start for Free* button.

3. Follow the onscreen instructions to create an account and start your free trial.

Configure G-Suite

If you use Google as your email host, you must configure it to work with Paubox.

4. Open a browser to *https://admin.google.com*.

5. Navigate to *Apps > G-Suite > Mail > Advance Settings > General Settings* tab *> Inbound Gateway*.

6. Select the *Edit* button for Inbound Gateway. Then configure as below, then tap *Save*:

Edit setting ✕

Inbound gateway   Help

Paubox   Edit

1. Gateway IPs

IP addresses / ranges   ADD

52.26.0.0/16

☐ Automatically detect external IP (recommended)
☐ Reject all mail not from gateway IPs
☑ Require TLS for connections from the email gateways listed above

2. Message Tagging

☐ Message is considered spam if the following header regexp matches

                                          CANCEL   SAVE

7. Scroll down near the bottom of the page to *Outbound Gateway*. Enter *outbound.paubox.com*

Routing

Outbound gateway   Route outgoing emails to the following SMTP server: ❓
Locally applied

                      outbound.paubox.com

8. At the bottom right corner of the browser, tap *Save*.

Configure DNS MX Records

9. Using a browser, log into your DNS records. In this example, it is GoDaddy.

10. Go to your *DNS records control panel*.

11. Add the following *MX* (Mail Exchanger) record:

- Host: *@*
- Points to: *mx2.paubox.com*
- Priority: *10*

12. *Save* the new MX record.

13. Delete any other MX records.

14. While still in your DNS records control panel, add the following *TXT* (Text) record:

    - Host: @

    - TXT Value: *v=spf1 include: _spf.paubox.com include: _spf.google.com -all*

15. *Save* the new TXT record.

16. Exit the DNS records control panel.

17. If you have been following along with the Paubox onscreen instructions, tap the *Next* button. This takes you to the *Confirm your DNS record settings* page.

Send email to a TLS Recipient

18. When sending email to a recipient who supports TLS encryption, all goes as normal, with the exception of an added footer stating *This email was seamlessly encrypted for your privacy and security by Paubox.*

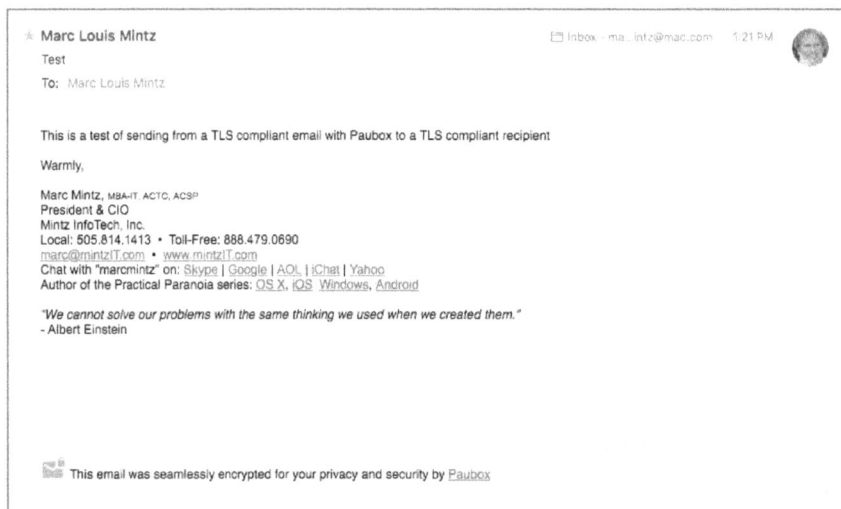

Send email to a non-TLS recipient

19. When sending email to a recipient who doesn't support TLS, your email is encrypted between your computer, to the Paubox mail server. Paubox then

sends an email to the recipient alerting them of the mail waiting at Paubox. The recipient taps the *View Message* button. A secure browser window opens to let the recipient view the email by tapping the *View Message* button. The recipient can reply to the email from this window securely with end-to-end TLS encryption.

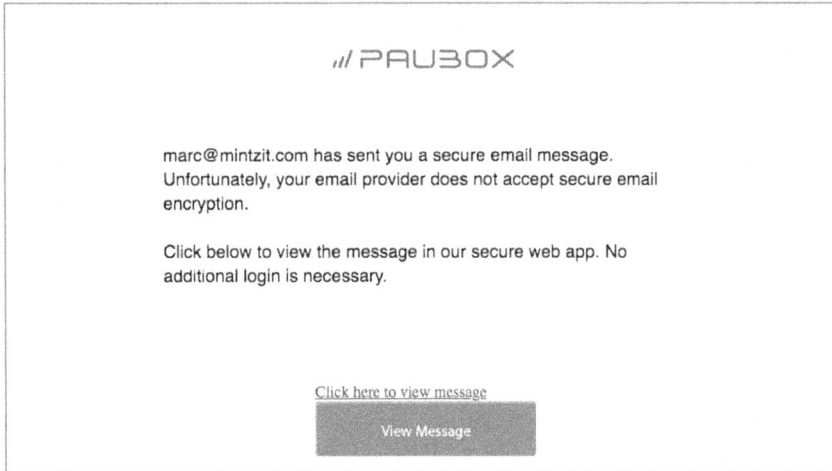

.ıl PAUƎOX

marc@mintzit.com has sent you a secure email message. Unfortunately, your email provider does not accept secure email encryption.

Click below to view the message in our secure web app. No additional login is necessary.

Click here to view message

View Message

Reply   (Login not required to reply)

To: marcmintz@mailinator.com

From: marc@mintzit.com

Subject: Test

Encrypted Message:

<<=======
This is a test of sending from a TLS compliant email with Paubox to a non-TLS compliant recipient

Warmly,

Marc Mintz, MBA-IT, ACTC, ACSP
President & CIO
Mintz InfoTech, Inc.
Local: 505.814.1413 • Toll-Free: 888.479.0690
marc@mintzIT.com • www.mintzIT.com
Chat with "marcmintz" on: Skype | Google | AOL | IChat | Yahoo
Author of the Practical Paranoia series: OS X, iOS, Windows, Android

"We cannot solve our problems with the same thinking we used when we created them."
- Albert Einstein

This email was seamlessly encrypted for your privacy and security by Paubox

13.5   HTTPS With Web Mail

We discussed HTTPS in the previous chapter. It is an encryption protocol used with web pages. It also can be used to secure email that is accessed via a web browser. When using HTTPS, your username and password are fully encrypted between your device and the email host server, as are the contents of all email that you create or open.

When using a web browser to access email, it is vital that your email site use the HTTPS encryption protocol to help ensure data and personal security.

13.5.1 Assignment: Configure Web Mail to Use HTTPS

If you use a web browser to access your email, it is critical that your web connection use HTTPS.

In this assignment, you verify that your browser-based email uses HTTPS.

- Note: If you do not use browser-based email, you may skip this assignment.

1. Launch your web browser.

2. Go to your login page for your email. In this example, we use Google Mail (Gmail).

3. As in the screen shot below, make sure that the URL field shows either the lock to the left of the URL, or *https://* and not *http://*. This indicates you are communicating over a secure, encrypted pathway.

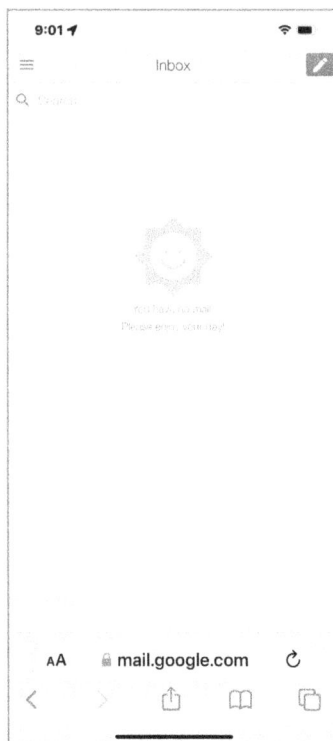

4. If instead, your browser shows the URL to be http://, try revisiting your email log in page, but this time manually enter `https://`.

5. If you get to the log in page, all is good. Just bookmark the https:// URL and use it instead of the previous non-secure URL.

6. If you cannot get to your log in page, change your email provider NOW!

13.6   Mail Privacy Protection

New with iOS 15 and macOS 12 is Mail Privacy Protection (MPP). With MPP active, your IP address is hidden from those sending you email, and tracking pixels are blocked. With tracking pixels blocked, email senders cannot determine your IP address if you have opened the email or gather any other information about you.

To accomplish MPP, you must opt in for the service, and use the built-in Mail.app. Once turned on, email is no longer routed directly to your computer. Instead, incoming email is routed through at least two proxy servers. The first thing the servers do is assign you a temporary IP address that is not linked to your machine or account, but is the IP address the sender sees. The IP address is associated with the geographical region you receive your email, but not your specific address.

The servers then watch for pixel trackers and then block their activity.

13.6.1   Assignment: Enable Mail Privacy Protection

In this assignment, you enable Mail Privacy Protection.

Prerequisite: iOS 15 or higher installed.

Prerequisite: Using Apple Mail.app

If Opening Mail.app For the First Time

1. Open *Mail.app*.

2. The *Mail Privacy Protection* window opens. Select *Protect Mail activity* radio button, and then tap *Continue* button:

If Mail.app Has Previously Been Opened

1. Open *Settings > Mail > Privacy Protection.* The *Privacy Protection* screen opens.

2. Turn on *Protect Mail Activity.*

3. Exit *Settings.*

13.7   Email Aliases and Hide My Email

Email Aliases

Almost all of us suffer from email overload. Although junk mail filters help, it is more band-aid than true solution. One of the fastest, easiest, and least expensive (as in *free*) solution I've found is to use a series of bogus email addresses, one for each sign up or registration you make on the web.

For example, if I'm interested in getting more product information from company xyz.com by filling out a web form, instead of using my real email address of *marc@thepracticalparanoid.com,* I may use *marc+xyz@thepracticalparanoid.com.*

Both email services that I use (Apple, with both a @mac.com and @icloud.com address, and Google, with both a @gmail.com and @thepracticalparanoid.com address) allow an unlimited number of email aliases by just adding *+sometext* between your account name and the @ sign. Your email provider may also offer this free feature.

All these aliases will be recognized as your base email account, but when receiving an email addressed to the alias, the alias will show in the email. This allows you to create rules within the mail application to manage the incoming mail. For example, should I later decide I no longer want to receive any communications from xyz.com, I can create a rule to automatically delete any incoming mail addressed to *marc+xyz@thepracticalparanoid.com.*

Below is an incoming email after subscribing for newsletters at thepracticalparanoid.com:

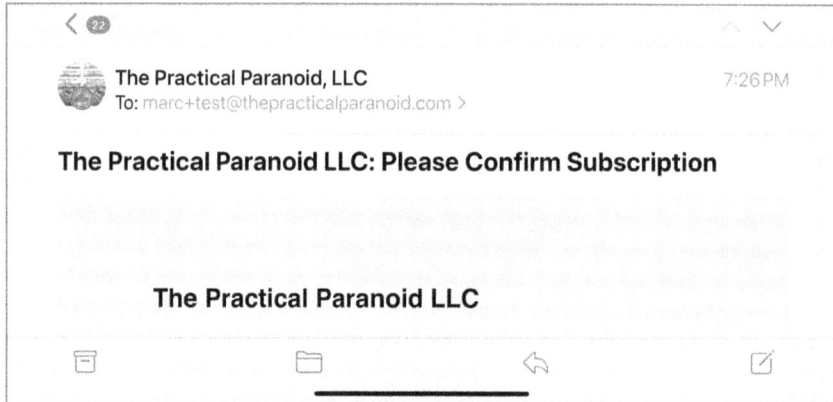

Hide My Email

New to iOS 15 and macOS 12 is *Hide My Email.* Hide My Email lets users share a unique, random email address that forwards to their personal inbox anytime they want to keep their personal email address private. This feature is built into Safari, iCloud settings, and the Mail app. It works only with Apple email accounts, such as *example@icloud.com.*

13.7.1   Assignment: Use an Email Alias

In this assignment, you determine if your email provider allows email aliases.

4. Open your email application.

5. Create a new email, addressed to an alias of your email account. For example, if your email account is *joe@email.com,* an alias could be *joe+test@email.com.*

6. Enter a subject in the subject line.

7. Send the email.

8. Wait a minute, then check for new email.

9. If your email provider allows for aliases, you will see the email in your inbox.

13.7.2   Assignment: Enable Hide My Email

In this assignment, you will enable Hide My Email.

● Prerequisite: iOS 15 and higher.

1. Open *Settings > your name > iCloud > Hide My Email.*

2. In the *Hide My Email* screen you can see one alias already created for you. If you wish, you can have additional names randomly created:

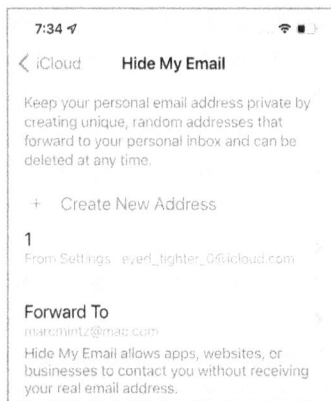

3. Select *Forward To* to select which of your Apple email accounts this alias is to be bound to.

4. When complete, tap *Done*.

5. Exit *Settings*.

13.7.3   Assignment: Use Hide My Mail

In this assignment, you will use your Find My Mail account to subscribe to a newsletter.

● Prerequisite: Completion of the previous assignment.

● Prerequisite: Use Safari.

30. Open Safari to a site that requires an email address from you. A site that wants you to subscribe to a newsletter is ideal.

31. Tap inside the *Email* field. The *Hide My Email* alert will pop-up. If you would like to use *Hide My Email,* tap on the alert.

32. The *Hide My Email* window opens.

    a. To use the new suggested address, tap *Continue* button.

    b. To use one of your existing addresses, tap *iCloud Settings > Apple ID >* press and hold on *Hide My Email* address to copy the address, return to the web page, then paste in the desired address.

13.8   End-To-End Secure Email With ProtonMail

If you are serious about email security, then you need to use an end-to-end secure email solution. Forcing TLS for incoming and outgoing email is one option (see previous section 13.5). However, it is likely that either the sender or the recipient use email hosts which do not allow the higher security, thereby forcing TLS actions.

There are two other options for point-to-point email encryption:

•   Use an email encryption utility. This works well if the other end of the communication also is using the same encryption utility. The next section covers this strategy using *GNU Privacy Guard* and *S/MIME*.

•   Use a cloud-based option. This method makes it every bit as simple to send and receive email as the user is accustomed to. The downside is that instead of using an email client, a website is used to send and receive mail. An example of this is *Sendinc.com*[9].

An interesting hybrid option is found in *ProtonMail*[10]. ProtonMail includes PGP public key/private key encryption, so that neither you nor the other party need deal with the potential headaches of installing and configuring PGP encryption.

---

[9] *https://sendinc.com/*

[10] *https://protonmail.com*

ProtonMail offers several advantages for the typical user, including:

- Free with optional monthly/yearly plans.

- Based in Switzerland so all user data is protected by Swiss privacy laws.

- Allows the user to determine the destruction date and includes unlimited retention.

- Allows for encrypted and password protected emailing to non-ProtonMail users.

- Allows for rich text email.

When sending from ProtonMail to a non-ProtonMail user, your recipient receives an email stating that a secure message is waiting. The recipient taps the link, taking the recipient to an authentication page. Upon entering the password, the recipient then sees the message. The recipient can directly and securely reply to the message, then you receive their reply in your inbox.

When sending from ProtonMail to ProtonMail, the interface is like other email providers.

Although not as convenient as using your own email software, we find ProtonMail to be an easy choice when security, convenience, and cost are taken into consideration against the impacts of data theft, or the potential drama of confidential communications being intercepted.

13.8.1   Assignment: Create a ProtonMail Account

In this assignment, you create a free ProtonMail account.

- Note: In this assignment you create a free account. This is limited and does not allow sending encrypted mail outside of ProtonMail. You can, however, upgrade the free account at any time.

1. Open a browser to visit *https://protonmail.com*.

2. Select the *Sign Up* button.

3. Scroll down to tap the dropdown arrow next to the plan you wish to use. In this tutorial, you create a free account. If you wish to use a monthly plan, make sure to double check the currency used on the bottom of the page.

4. Tap the *Select Free Plan* button.

5. Enter the *Username* and *Password* you wish to use. We recommend using easy to remember 15-character passphrases.

6. Enter a *Recovery email*, so you can regain control over your account if locked out or forget your password.

7. Select Create Account.

8. Follow the onscreen instructions to authenticate your identity.

9. ProtonMail begins to create your account.

10. At this stage enter the name that is to be seen by other users. You also have the option of downloading iOS or Android Apps. Next tap on the *Finish* button.

11. You have now finished the setup process. You see a short tutorial on the bottom of your screen, it is recommended to read through it to understand some more of the features available to you.

13.8.2   Assignment: Create and Send an Encrypted ProtonMail Email

In this assignment, you send your first fully encrypted email through ProtonMail.

- Prerequisite: An existing ProtonMail account, or completion of the previous assignment.

1. Open a browser to *ProtonMail* at *https://ProtonMail.com*, select the *Login* link, then log in.

2. Tap *New Message* button. The *New Message* window opens.

3. Scroll to the bottom of the page, then tap the lock (encryption) icon. This allows you to set a password requirement to open the email from a non-ProtonMail account. Configure to your taste, then tap the *Set* button.

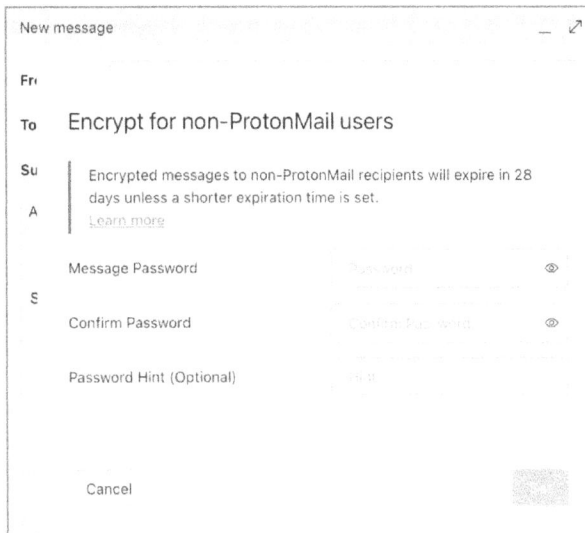

4. Finish configuring your email, then tap the Send button. The program takes a moment to encrypt then send.

Notification of your email has been sent to the recipient.

13.8.3   Assignment: Receive and Respond to a ProtonMail Email

In this assignment, you reply to a ProtonMail secure email.

- Prerequisites: Completion of the previous two assignments.
- Note: For this assignment, screenshots from a computer instead of iOS are used for clarity.

1. If you have sent an email to a non-Proton Mail account, and have not encrypted with a password, the recipient receives an email no different from any other account.

2. If you have sent an email to a non-Proton Mail account, and have encrypted with a password, the recipient receives an email notice. To view the message, the recipient selects the *Unlock Message* button within the email.

3. After entering the required password, the email is displayed in the recipient's browser. The recipient is also able to reply via this webpage by selecting *Reply Securely.*

13.9   [macOS and Windows] Outlook.com Encryption and Prevent Forwarding

13.9.1   [macOS and Windows] Send an Encrypted Email From Outlook

13.9.2   [macOS and Windows] Read an Encrypted Email From Outlook.com

13.10 PGP/GPG & S/MIME

Until recently, the gold standards for email security were PGP, GNU Privacy Guard, and S/MIME. Unfortunately, configuring each is only a few steps less technically intricate than rocket science, both parties must have the same software installed, and S/MIME requires the purchase of yearly certificates by each party.

For these reasons I do not generally recommend use of these security protocols. With just 2 minutes of form entry, one can have a ProtonMail account with preconfigured PGP.

13.11 Email Validation With SPF, DKIM, and DMARC

Sender Policy Framework (SPF)

SPF[11] is an email-validation system. It provides a mechanism to authorize servers and services to send email using your domain. This allows a receiving mail exchanger (mail server) to verify that incoming mail from a specific domain is coming from a host authorized to send that mail. When a criminal hacker sends email to you with fake "from" information (for example, the sender pretends to be

---

[11] *https://en.wikipedia.org/wiki/Sender_Policy_Framework*

a vendor submitting an invoice for payment), your email server can validate or invalidate the sender as authentic. The purpose of this SPF is to prevent email with forged addresses from reaching an inbox.

If the sender of an email is validated, the email comes on through as it always has. If the sender is invalidated, this tells the receiving server to follow the rules laid out in the DMARC record (explanation below after DKIM topic) when dealing with the unauthorized email.

Domain Keys Identified Mail (DKIM)

DKIM[12] is an email authentication system to detect spoofing. It provides a mechanism for the receiver to verify that an email stating to have come from a server which has been authorized to send mail for a specific domain via SPF record, is indeed the server that is sending the email. This is done via security key exchange, the public part of which is a 2048 or larger sha256 hash algorithm encrypted key, which you will place into your DNS records in the next assignment. The DKIM purpose is to prevent acceptance of emails sent from spoofed services and servers.

DKIM works by attaching a digital signature to each outgoing email. The recipient's email system validates the signature. These signatures are normally not visible to the user.

Domain-based Message Authentication, Reporting & Conformance (DMARC)

DMARC[13] is the configurable policy detailing how to deal with email that has failed the SPF and/or the DKIM validation. The easiest way to think about the DMARC process is as follows.

- SPF authorizes a server to send mail on behalf of a specific domain.
- DKIM authenticates the sending server truly is the authorized one.
- DMARC defines what to do with the email when it fails.

The options are *take no action, quarantine the email,* or *reject the email.*

---

[12] *https://en.wikipedia.org/wiki/DomainKeys_Identified_Mail*
[13] *https://en.wikipedia.org/wiki/dmarc*

13.11.1 Assignment: Configure SPF

In this assignment, you configure SPF for your email domain.

- Prerequisite: Creating SPF records is only possible if your email is on its own Fully Qualified Domain Name (such as *thepracticalparanoid.com*). If you are using a public domain such as *gmail.com, xfinity.com, myschool.edu*, you do not have the ability to edit your DNS records, and therefore cannot create your own SPF records.

- Prerequisite: Administrative access to your domain DNS records and control panel.

In this assignment, I use the domain *thepracticalparanoid.com* which is hosted with Google G-Suite, with DNS hosting at *GoDaddy.com* as the example. If your email or DNS hosts are different, the necessary steps may differ as well.

1. Open a web browser to your DNS Control Panel at your DNS host.

2. Select *Edit*.

3. Create a new *TXT* record with the following values:

   o For *Name/Host/Alias* enter @

   o For *Time to Live* enter 3600

   o For *Value/Answer/Destination* enter
      v-spf1 include:_spf.google.com ~all

      ▪ Note: Check with your email vendor technical support for the proper Value/Answer/Destination to replace my string.

Type *	Host *		TXT Value *
TXT	⇕	@	v-spf1 include:_spf.google.com
TTL *			
1 Hour ⇕			
			Save Cancel

4. *Save* the DNS changes.

5. *Verify* the DNS changes. This is done in Google from *https://toolbox.googleapps.com/apps/checkmx/*

Google Admin Toolbox Check MX

Domain name

thepracticalparanoid.com| RUN CHECKS!

DKIM selector (optional)

6. Enter your domain name, then select *Run Checks!*

   o Note: Some DNS servers are not configured to send all necessary records through Google. If you are certain you have entered all your DNS records correctly and still receive an error when using the Google Admin Toolbox, try using https://mxtoolbox.com instead for steps 5 and 6.

7. For a detailed description of the SPF syntax, the service dmarcian.com has the most detailed, yet still user-friendly guide to understanding SPF: *https://dmarcian.com/spf-syntax-table/.*

8. If any errors are reported, discuss them with your DNS host support staff to have them resolved.

13.11.2 Assignment: Configure DKIM

- Note: Creating DKIM records only is possible if your email is on its own Fully Qualified Domain Name (such as *thepracticalparanoid.com*). If you are using a public domain such as *gmail.com*, you do not have the ability to edit your DNS records, and therefore cannot create your own DKIM records.

In this assignment, I use as the example the domain *thepracticalparanoid.com* which is hosted with Google G-Suite, with DNS hosting at GoDaddy.com. If your email or DNS hosts are different, the necessary steps may differ as well.

Generate a public domain key for your domain

1. Open a browser to *admin.google.com.*

2. Select *Apps > G-Suite > Gmail > Authenticate email.*

3. Select the target domain for which you want to generate a domain key.

4. Select *Generate New Record*.

5. Select *Generate*.

6. A text box opens to display a 2048-bit key.

7. Select, then copy, this key.

Create a DKIM record

8. Open a new web page, then go to your *DNS Control Panel* at your DNS host.

9. Select *Create a new TXT* record.

10. In the *TXT Value* field, *paste* in the key created in step 6 above.

11. In the Host field, enter `google. _domainkey`.

TXT		
Host *	TXT Value *	TTL *
google._domainkey	v=DKIM1; k=rsa; p=MIIBIjANBg	1 Hour

Save   Cancel

12. Save the changes made to your DNS records.

13.11.3 Assignment: Sign Email with The Domain Key

In this assignment, you configure your mail server to automatically attach the DKIM key to all outgoing email.

• Prerequisite: Completion of the previous assignment.

1. Open a browser to *admin.google.com*.

2. Select *Apps > G-Suite > Gmail > Authenticate email*.

3. Select the target domain for which you want to attach a domain key.

4. Select *Start authentication*.

13.11.4 Assignment: Configure DMARC

Once DKIM is in place, a decision must be made what to do with incoming email found to be spoofed or fake. A general recommendation is this:

1. Configure DMARC to do nothing with failed validations, and to notify the administrator. When first instituting the DMARC record, use this setting to observe all authorized and unauthorized email sent over the course of a few weeks. This helps to ensure you have the correct records in place and no false positives are found.

2. Reconfigure DMARC to place failed validations in quarantine and to notify the administrator. Leave on this setting for a week or two. If false positives are found in quarantine, research the reason and resolve. Once false positives have been eliminated, go to the next setting.

3. Reconfigure DMARC to reject failed validations.

In this assignment, you configure your mail server to do nothing with failed validations and to notify the administrator.

1. Open a browser to your DNS control panel at your DNS host.

2. Create a new TXT record with the following attributes:

- For Record Name/Host enter is `DMARC`

- For *Value* enter on one line:
  v-DMARC1; p=none; rua=mailto:webmaster@thepracticalparanoid.com

  - Substitute your administrator email address in place of *webmaster@thepracticalparanoid.com*.

  - To send failed validations to quarantine, substitute `p=quarantine`.

  - To reject failed validations, substitute `p=reject`.

TXT			
Host *	**TXT Value** *	**TTL** *	
_dmarc	v=DMARC1; p=quarantine; rua=	1 Hour	⬍
		Save	Cancel

3. Save your changes.

4. To test your DMARC record, go to *https://dmarcian.com/dmarc-inspector/*

Refer to the *DMARC Tag Registry*[14] for other available options.

13.12 2-Factor Authentication

As with all your other important online accounts, it is mandatory that each of your email accounts be secured with 2-Factor Authentication.

If you have an email account that does not permit 2FA, *run,* don't walk, away from that email service.

13.13 Email Lessons Learned

☐ Phishing is the attempt to acquire your sensitive information by appearing as a trustworthy source, usually via email.

☐ Secure Socket Layer (SSL) is the original email encryption protocol, now depreciated due to vulnerabilities.

☐ Transport Layer Security (TLS) is the successor to SSL and the current email encryption standard.

☐ Hypertext Transport Layer Secure (HTTPS) is the current encryption protocol for browsing.

☐ New to iOS 15 and macOS 12 is Mail Privacy Protection, which can help secure email against pixel tracking and other vulnerabilities.

☐ New to iOS 15 and macOS 15 is Hide My Email, which creates on-the-fly email aliases linked to your @mac.com or @icloud.com accounts.

☐ Email aliases can be created in some email systems (such as @mac.com, @icloud.com, and @gmail.com) on-the-fly by adding "+aliasname" between your email name and the @ symbol.

---

[14] *https://dmarc.org//draft-dmarc-base-00-01.html#iana_dmarc_tags*

☐ By using checktls.com you can verify if both the sender and recipient can use TLS for secure email communications.

☐ By using Paubox you can force encrypted email communications, even if the other party does not support it.

☐ ProtonMail provides for secure encrypted email using built-in GNU Privacy Guard and S/MIME.

☐ Sender Policy Framework (SPF) is an email-validation system allowing receiving mail exchangers to verify that incoming email is coming from a host authorized to do so for that domain.

☐ Domain Keys Identified Mail (DKIM) is an email authentication system to detect spoofing.

☐ Domain-based Message, Authentication, Reporting & Conformance (DMARC) is a configurable policy detailing how to deal with email that has failed the DKIM validation.

☐ Enabling 2-Factor Authentication in your email is a vital foundation to email security and privacy. If your email provider does not allow for 2FA move to another provider.

14 Documents

No matter how paranoid or conspiracy-minded you are, what the government is actually doing is worse than you imagine.

–William Blum[1], American author, and former State Department employee

What You Will Learn in This Chapter

- Use file encryption
- View EXIF data in pictures
- Remove EXIF data in pictures
- View and edit metadata in MS Office files
- View and edit metadata in pdf files
- Redact data in pdf files

What You Will Need in This Chapter

- [Optional] $10.99/year subscription for Fox IT PDF.
- [Optional] $9.99/year subscription for WinZip.
- Digital photograph in .jpg or .tiff format.

14.1 Document Security

---

[1] *https://en.wikiquote.org/wiki/William_Blum*

If your documents never leave your device, and you have encrypted your storage devices using FileVault 2, there is no need to go the extra step to encrypt your documents. But should you ever need to email your sensitive data to someone else, or pass a sensitive data via any storage device, encrypting the data goes a long way to ensuring your peace of mind.

There are several options to document encryption, each with its own benefits and drawbacks. We discuss each here.

14.2   Password Protect a Document Within its Application

A few applications are designed with document security in mind and offer their own encryption schemes. Microsoft Office and Adobe Acrobat Pro are common examples.

Although Microsoft Office products make it an easy process to password protect your documents, prior to Office 2007 (Windows) and 2011 (Mac), it was an equally easy process to break the encryption. There are many freeware and commercial utilities that can bypass the password in older versions and open the document for reading.

Starting with Microsoft Office 2007 and 2011, Microsoft changed its encryption standard to use the secure AES-128[2] algorithm. Microsoft Office 2016 (Office 365) uses AES-256[3]. Assuming an adequate password length has been selected, some researchers estimate that it would take millions of years to brute-force crack an AES-256 password with current computing power. For security during this lifetime (famous last words), the AES-256 encryption standard should be enough to protect your documents if a strong password has been chosen.

- Note: As of this writing, only the macOS and Windows desktop versions of any Microsoft Office software can password-protect files or open password-

---

[2] *https://en.wikipedia.org/wiki/Advanced_Encryption_Standard*

[3] *https://technet.microsoft.com/en-us/library/cc179125%28v=office.16%29.aspx?f=255&MSPPError=-2147217396 - About*

protected files. The online, Android, and iOS versions cannot do either of these actions.

14.2.1 Assignment: Encrypt an MS Word Document

Although Microsoft Office products for the desktop and Web offer AES 256-bit encryption, same is not available to iOS versions.

If you need to encrypt a MS Office file, one option is to open it in Apple Pages, which does have encryption options.

See the next Assignment.

14.2.2 Encrypt an Apple Pages Document

In this assignment, you encrypt an Apple Pages file from within the application.

- Prerequisite: Apple Pages app must be installed on the iOS device.

1. Open Pages.

2. Tap the *Create New* icon.

3. From the *Templates* screen, select *Blank*.

4. Type a few characters, then tap the *More* (…) icon.

5. In the *More* screen, scroll down then tap *Set Password*.

6. In the *Set Password* screen, enter the desired strong password in the *Password* field, re-enter it in the *Verify* field, then tap *Done.* You are returned to your document.

7. You may continue to edit the file.

8. When ready to save, tap *Done,* then tap <. Your document is saved, and you are taken to the *Pages Recents* screen. The default name for a new document is *Blank.* Tap on the name *Blank.*

9. In the *Rename Document* screen, enter the desired name for the document. For this assignment, that is *First Encrypted File*, then tap *Done.*

10. At the Home screen, tap the *First Encrypted File.*

11. If you had enabled Touch ID for this file, at the prompt, touch the home button to authenticate opening the file. Otherwise, at the *Password* screen, enter the password, then tap *Done.*

The encrypted, password-protected file opens for editing.

14.3   Encrypt a PDF Document

As there are only a few applications that can encrypt their own documents, chances are you will work with a file whose application cannot perform the encryption. iOS can "print" any document to pdf format, and in the process, add password-protected encryption to the pdf using third-party software.

14.3.1   Assignment: Convert a Document to PDF for Password Protection

In this assignment, you convert a file, photo, or website to PDF.

- Note: As of iOS 16 dated 20221226, this function is not working.

1. Open the target file, photo, or web page web page.

2. Tap the *Share* icon > *Print.*

3. In the *Print Options* area, all pages on the site display. Tap to select the desired pages to be saved.

4. Zoom (spread fingers) on a page to be printed. The pages display full screen.

5. Tap the *Share* icon.

6. From the *Share* screen, you can send the pages in pdf by any means available to your device. For this assignment, tap *Save to Files > Documents.*

7. Tap on the default name to rename, then tap *Done.*

8. Tap Save.

The file is now saved as a pdf in your device *Documents* folder or cloud service.

14.3.2 [Windows] Assignment: Convert an MS Word Document to PDF for Password Protection

14.3.3 [Optional] Assignment: Encrypt a PDF

iOS currently has no native ability to encrypt a PDF, but there are several utilities that can. Although Adobe Acrobat is free for iOS, to enable its encryption service requires a $9.99/month subscription. FoxIT PDF has the same feature for $10.99/year subscription.

In this assignment, you convert a file into a PDF for encryption using FoxIT PDF.

- Note: Requires installing FoxIT PDF and signing up for a free trial subscription.

Install Foxit PDF

1. Open *App Store.app.*
2. Search for then download FoxIT PDF.

Sign Up for Trial Subscription

3. Open *Foxit PDF.*
4. Tap the *3 Lines* (menu) icon > *Sign In* > *Login/Sign Up* button.
5. In the *FoxIT ID* screen, tap *Sign up for a Foxit ID.*
6. Enter your *Email,* a *Password,* and your *Name,* then tap *Sign Up.*
7. An email with a verification link is sent to you. Open the email, then tap the verification link.
8. Return to FoxIT PDF.
9. Tap the *3 Lines* (menu) icon > *My Account.*
10. Tap *Subscribe* > *$10.99/year,* then follow the onscreen instructions to authorize.

Configure Foxit for Cloud Service Access

The fastest/easiest way to give FoxIT access to your files is through a cloud service such as Google Drive, Dropbox, Apple iCloud, and Microsoft OneDrive.

11. Open FoxIT PDF.

12. Tap the *3 Lines* icon > *Cloud.*

13. Tap *+Add.*

14. Select the cloud service to use.

15. Enter your *username* and *password* for the service, then tap *Authorize.* You may need to follow the onscreen instructions to complete the connection.

Open File in Foxit PDF

- Note: As of this writing, FoxIT PDF Mobile has a file size limit of 5MB.

16. Open Foxit PDF.

17. Tap the *3 Lines* icon > *Cloud* > *target cloud service* > navigate to select the target file.

18. When the file opens in Foxit PDF, tap the *Convert* icon.

19. In the *Convert to PDF* screen, tap *Convert.* The file converts to PDF and open in Foxit PDF.

Encrypt a PDF

20. With the PDF file open in Foxit PDF, tap the *3 dots* > *File Encryption.*

21. Select the desired security options to be protected by password, configure to your taste. When complete, tap *Done*:

22. Open Document.

23. Add Document Restrictions.

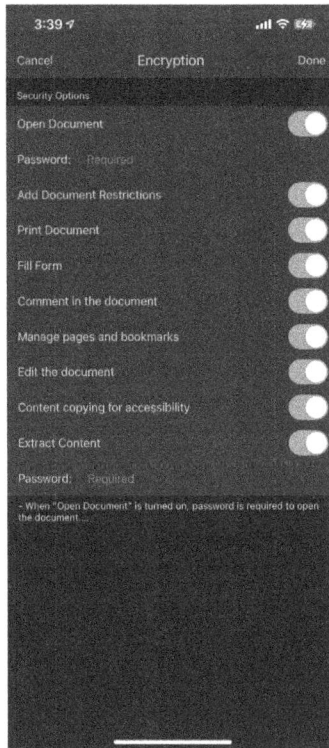

24. Save the encrypted PDF to your desired location.

14.4   [macOS and Windows] Encrypt a Folder

14.4.1   [macOS] Assignment: Create an Encrypted Disk image

14.4.2   [Windows] Assignment: Create an Encrypted Windows Folder

14.4.3   [Windows] Assignment: Backup Encryption Certificate for the Encrypted Folder

14.4.4 [Windows] Assignment: Import Previously Saved File Encryption Certificates

14.5 Encrypt a File or Folder for Cross Platform Use with Zip

If you need to exchange a file or files with others regardless of the computing platform used, password protect the archive in a format that is readable by any OS. Although there are over a dozen cross-platform compression formats, *zip* has become the most common standard.

iOS does not have the built-in ability to create zip archives, but there are several third-party utilities available. I recommend WinZip.

Once you have created an encrypted archive of your file or files, the archive can be uploaded to a file server, shared by email, or passed along via drive, disc, or thumb drive.

- Note: The encryption protocol used in *zip* is considered weak and should not be used for HIPAA, SEC, legal, or other high-security needs unless using a third-party zip utility that provides AES 256-bit encryption. *WinZIP* is the industry leader for cross-platform commercial software that provides this level of *zip* security. *7-zip* is the industry leader for open-source software with this level of zip security. *Keka* uses *7-zip* as well as *zip with AES 256*.

14.5.1 Assignment: Encrypt a File or Folder Using Zip

In this assignment, you create a folder in which you copy a file or photo, then encrypt the folder and its contents using the commercial *WinZip Pro* app.

- Note: This assignment requires the purchase and download of WinZip (commercial software) from the Apple App Store. Free 3-day trial, then $9.99/year subscription.

Install WinZip Pro

1. On your iOS device, open the *App Store* app.

2. Tap the *Search* button, then search for *WinZip Pro.*

3. Select *WinZip Pro,* tap the *Download* icon, agree to pay, then download *WinZip Pro.*

Configure WinZip preferences

4. Open *WinZip.*

5. Tap *Settings* icon (gear).

6. In the *Settings* screen, configure as follows:

- In the *Cloud Service* area, if you have an account with any of the offered services (Dropbox, Google Drive, OneDrive, or iCloud), tap to sign in and link WinZip to the service.

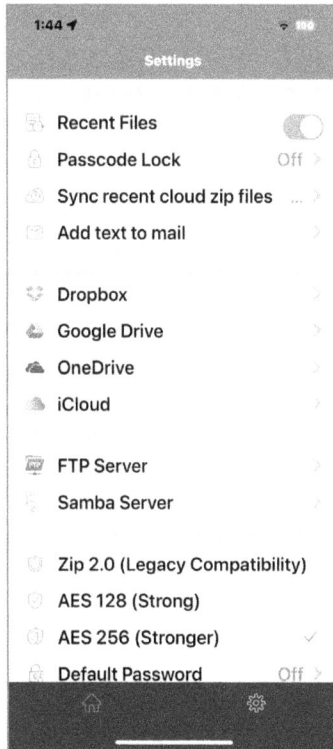

- In the *Encryption* area, enable *AES 256 (Stronger).* This forces WinZip to use the most modern and secure version of the zip format.

7. Tap the WinZip *Home* button.

Create a new folder within WinZip Pro

8. Open WinZip.

9. In the WinZip Home screen, tap My Files.

10. Tap the *Edit* icon (pencil), tap *New Folder* button, name the new folder *Test*, then tap *Create*. The new folder appears in the *My Files* screen.

11. Tap *Done*, then tap *WinZip* to return to the home screen.

Copy a file into the WinZip folder

12. From the WinZip home screen, tap *Photos*. This takes you to your iOS Photos library.

13. Navigate through your photos to locate a target image. Tap the image. It opens.

14. Tap the *More* icon > *Copy*.

15. Tap *My Files* > *Test,* then tap *Paste.* The photo is now copied into the *test* folder.

Encrypt the WinZip folder

16. Return to the WinZip home screen, then tap *My Files*.

17. Tap the *Expand* icon to the far right of the folder name > *Zip* > *Zip Here*.

18. In the *Zip File* dialog, you may rename the new zip archive, encrypt with a password (for this assignment use password), then tap *OK*.

19. Your new password-protected encrypted archive is now available to email.

Email the zipped archive

20. In WinZip home, navigate to the archive to be emailed. For this assignment, tap *My Files*.

21. Tap the Edit icon (pencil), then enable the radio button for the target archive. For this assignment, *Test.zip*.

22. Tap *Send Mail* button.

23. The *Mail* app opens, with the target file attached. All that is left to do is address the email, enter a subject line, then tap *Send*.

14.5.2   [macOS] Assignment: Set Keka as the Default for Zip

14.5.3   Assignment: Open an Encrypted Zip Archive

In this assignment, you open the encrypted zip archive created in the previous assignment.

- Prerequisite: Completion of the previous assignment.
1. Double tap the encrypted zip archive.
2. The archive opens into WinZip Pro.
3. At the prompt, enter the password to unzip and open the file.

14.6   Exchangeable Image File Format and Metadata

The Exchangeable image file format[4] (Exif) is a standard for storing metadata for image (JPEG, TIFF, WAV, but not JPEG 2000 or GIF) and sound files. *Metadata*[5] is "information about the information." Typically, the metadata is stored within the file itself, but the standard does allow for the Exif data to be stored separately from the file.

The following image data may be stored:

- Camera settings: aperture, focal length, ISO speed, metering mode, and shutter speed
- Copyright information
- Creation date and time
- GPS (Global Positioning System) coordinates
- Image thumbnail
- File description.

---

[4] *https://en.wikipedia.org/wiki/Exif*
[5] *https://en.wikipedia.org/wiki/Metadata*

Developers may opt to include additional image information.

The following audio data may be stored:

- Bits per sample
- Bytes per second (average)
- Date createdEncoding format
- Exif version
- Make
- MakerNote
- Model
- Number of channels
- Related image file
- Sampling rate
- Time created

Developers may opt to include additional audio information.

The issue with Exif is the embedded GPS data. This value can place the exact location on earth where the photograph was taken. Add the creation date and time and there exists a 4-dimensional positioning for where and when the photograph was taken.

As an example of how this affects security and privacy, let's say you take a picture of your fine sushi dinner and post it. As most websites and social media do not strip out Exif data, anyone viewing the image can know that, based on the date/time stamp and GPS, you took the photo 5 minutes ago at a restaurant 100 miles from your home. A couple of hours allow more than enough time to rob your home and get away.

14.6.1   Assignment: View Exif Data in a Photograph

In this assignment, you view the Exif data in a photograph.

- Prerequisite: A digital photograph in JPG, TIFF, or WAV format on your computer. If you do not have a file in this format, take a picture with your phone or tablet.

Use Exif Metadata to view Exif

iOS has no built-in utility to view or edit Exif data. This requires a 3rd-party tool. There are many available on the Apple App Store. For this assignment, you are using *Exif Metadata*.

1. Open the *App Store* app, search for then download the free *Exif Metadata* app.

2. Launch Exif Metadata.

3. On the Exif Metadata home screen, tap the + icon to select a photo.

4. In the *Photo Albums* screen, tap *All Photos,* then tap the target photograph. The image opens, displaying a scrolling list of Exif data.

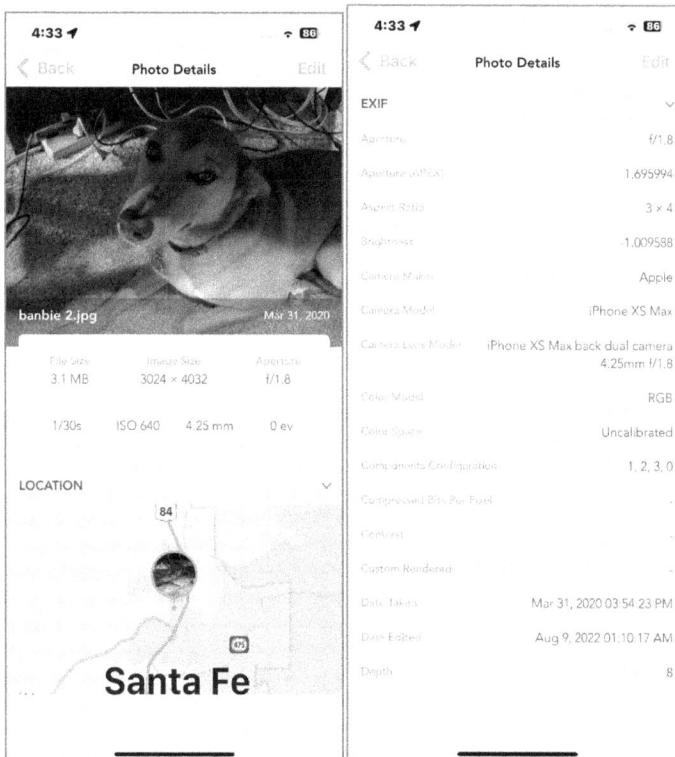

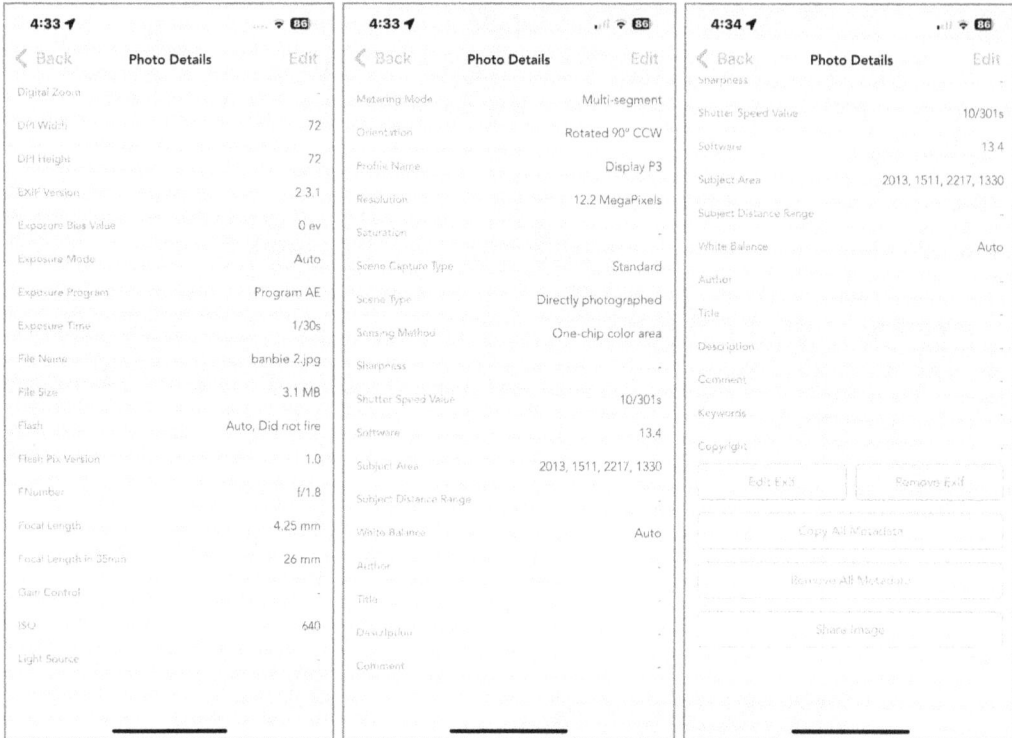

14.6.2 Assignment: Edit and Delete Exif Data

GPS location is one of many data points that can be isolated from Exif and affect your privacy. Before you distribute images, you can remove some or all your Exif data.

Unfortunately, iOS does not include the tools to do this, but there are many free and for-fee utilities that can. My favorite is *Exif Metadata*.

In this assignment, you edit or delete Exif data from your image.

- Prerequisite: A digital photograph.

Edit Exif Metadata

Exif Metadata allows you to edit some of the Exif metadata.

242

1. While in Exif Metadata, select the target image, scroll to the bottom of the *Photo Details* screen, then tap *Edit Exif.* The *Edit Exif* screen opens.

2. To edit an Exif data record, tap on the record, enter the desired data, tap *Save,* then tap *Cancel* to return to the *Photo Details* screen.

Edit Location

3. In the *Photo Details* screen, scroll down to just below the GPS map, tap *Edit Location,* drag the pin to the desired location, tap *Save,* then tap *Cancel* to return to the *Photo Details* screen.

Delete Location

4. In the *Photo Details* screen, scroll down to just below the GPS map, tap *Remove Location.*

5. At the verification alert to delete the original file, select either *No* or *Yes,* then tap *Cancel* to return to the *Photo Details* screen.

Delete Exif Metadata

Exif Metadata allows you remove much of the Exif data.

6. At the bottom of the *Edit Exif* screen, tap *Remove Exif.*

7. An alert appears: Exif metadata has been successfully deleted and a copy of this photo has been saved to your library. Do you want to delete the original file? Select either No or Yes.

• Note: In the current version, GPS data remains with the copy.

8. Return to the Finder.

9. Select the edited image, then select *File* menu > *Get Info.* Note that the GPS data no longer displays.

10. Return to ApolloOne.

11. With the same image select in ApolloOne, select *Exif Toolbox* > *Remove All Data.*

12. In ApolloOne, in the *Inspector* window, view all the remaining Exif data.

• Note: While most identifying information is now missing, there remains plenty of data about the composition of the file itself. Different Exif editing utilities give differing control over the Exif data.

14.6.3 [Optional] Assignment: View Metadata from an Office 365 File

Microsoft Office is one of the most common file formats for exchanging information. All files hold metadata that may be considered sensitive or private. In the case of legal documents and medical records, users may encounter lawsuits and fines for including metadata in these files.

Unfortunately, as of this writing, although this can be done in macOS and Windows, there is no way to remove this metadata from within iOS. You can, however, view your primary metadata.

In this assignment, you view metadata from a Microsoft Office 365 files.

- Prerequisite: Access to a Microsoft Office 365 file, and installation of Microsoft Word, Excel, or PowerPoint for iOS.

Install Microsoft Word for iOS

1. If you do not currently have Microsoft Word for iOS installed on your device, open *App Store.app.*

2. Search for then download *Microsoft Word.*

3. Open Microsoft Word.

4. Link to your Microsoft Account.

View Microsoft Word File Properties

5. Open the target file into Word.

6. Tap the *3 dots > Properties.* The *Properties* screen appears. When done reviewing the document properties, tap *Back,* then *Done* to return to the file.

14.6.4 [macOS and Windows] Remove Metadata from an Office 365 Excel or PowerPoint File

14.6.5 [Optional] Assignment: Redact Content in PDF

If you have sensitive data within a pdf file, FoxIT PDF provides a tool to redact (black out) and remove this data.

* Prerequisites: Purchase and installation of FoxIT PDF and a pdf file.

1. Open your pdf in FoxIT PDF.

2. Tap *3 Dots > Protect > Redaction*. The file opens in the *Redaction* screen.

3. At the bottom of the screen you will see three icons: *Redact Text, Redact Pages,* and *Search.*

4. Tap the *Redact Text* tool, then highlight the text to be blacked out.

5. Tap *Apply All*. All highlighted text is now permanently blacked out and scrubbed from the file.

14.6.6 [Optional] Assignment: View Object Metadata in PDF

Metadata in a PDF takes three forms:

- *Properties.* This includes file name, author name, subject, and keywords.

- *Object metadata.* This is metadata embedded in a file that is embedded in the pdf, typically a photograph inside a text document.

- *File revisions.* This includes changes to the text of the file.

If an object saved in a pdf has metadata attached to it, the macOS and Windows desktop versions of Adobe Acrobat Reader and Adobe Acrobat Pro allow you to see the metadata.

- Prerequisites: Installation of Adobe Acrobat Reader or Adobe Acrobat Pro on a macOS or Windows computer, and a pdf file containing an object with metadata.

1. Open your pdf in *Adobe Acrobat Pro.*

2. From *Tools,* select *Edit PDF > Edit Text & Images.*

3. Tap on an object, then right-tap the selection > *Show Metadata.* If metadata is available, it displays. If no metadata is available, the *Show Metadata* submenu does not display.

14.6.7   Assignment: View and Edit Document Metadata in PDF

In this assignment, you read the document metadata of a PDF file and change metadata.

Although Acrobat Reader on macOS and Windows allows viewing and editing the pdf document properties, and Acrobat Pro on macOS and Windows allows for viewing object metadata, iOS does not yet have access to these tools.

1. If you do not already have Adobe Reader installed, open a browser to *https://adobe.com,* then download Adobe Reader.

2. Open a browser to *https://2pdf.com.*

3. Tap the *Edit PDF metadata* icon.

4. Tap the *Choose file* button.

5. Navigate to locate your target pdf.

6. The 2pdf.com page displays the current document metadata.

7. Edit the metadata to your taste.

8. Tap the *Update PDF Metadata* button.

9. Tap the *Download* button. The pdf document downloads to your iOS device with the edited or deleted metadata.

10. As you are not able to view metadata on iOS, tap the *Send* icon at the bottom of the screen, then email the pdf to yourself.

11. Open the emailed pdf on a computer with Adobe Acrobat Reader.

12. In Acrobat Reader, select *File* menu > *Properties* > *Description.* The document metadata will display.

13. Note how the metadata now displays with your modified data.

14. Quit Adobe Reader.

14.6.8 [macOS and Windows] Assignment: Remove All Metadata in PDF

14.7 [Android] Secure Folder

14.8 Documents Lessons Learned

☐ Microsoft Office 365 macOS and Windows desktop versions can encrypt its files using AES-256. This is the current industry standard. Earlier versions used vulnerable encryption.

☐ Apple *Pages* on iOS can create encrypted documents.

☐ Most file formats can be converted to pdf using the *Share* > *Print* command.

☐ iOS does not have built-in tools to encrypt PDF files, but Foxit PDF with an in-app upgrade to Pro subscription can do this.

☐ iOS does not have built-in tools to compress files and folders using *zip*, but *WinZip Pro* can do this.

☐ Exchangeable image file format (Exif) is a standard for storing metadata for jpeg, tiff, wav image files and sound files.

☐ iOS does not have built-in tools to view or edit Exif data, but the freeware *Exif Metadata* can do this.

☐ Microsoft Office document metadata can be viewed on iOS using *Microsoft Word, PowerPoint,* or *Excel* for iOS.

☐ PDF files may contain three types of metadata: properties, object metadata, and file revisions.

15 Voice, Video, and Instant Message Communications

Surveillance technologies now available–including the monitoring of virtually all digital information–have advanced to the point where much of the essential apparatus of a police state is already in place.

- Al Gore[1]

What You Will Learn in This Chapter

- Install and configure *Signal*
- Secure instant message with *Signal*
- Secure voice and video calls with *Signal*

What You Will Need in This Chapter

- No additional resources required.

15.1 Voice, Video, and Instant Messaging

Every time you send or receive a text message, phone call, or videoconference on your computer or mobile device, your conversations and metadata are stored by third parties. The manufacturers or developers (such as Apple, Facebook, Google, etc.) and carriers (Verizon, AT&T, etc.) for each party can intercept any traffic that crosses their networks. This interception may extend to any third parties that

---

[1] *https://en.wikipedia.org/wiki/Al_Gore*

work with your carrier, such as contractors, or subsidiaries. In addition, your local, state, and federal government monitor data in dragnet-style snooping.

How can you communicate easily and securely?

If you are interested in cross-platform, end-to-end encrypted, text, voice and video conferencing solutions, a few options are available.

New to macOS 12 and iOS 15 is Apple's FaceTime messaging service on Android and web, allowing Windows and other users share in secure messaging with the rest of us. Although Facetime is encrypted, Apple apparently does have some limited back door access. Because of this, we can only give FaceTime a 9.5 out of 10 in the cybersecurity rating.

Wire[2] and *Signal*[3] are our choices for end-to-end encrypted voice, video, instant messaging, and group communications. Both provide end-to-end encrypted communications between Android, Chrome OS, iOS, macOS, and Windows.

Wire is a for-fee commercial service. It offers a free 30-day trial.

Signal is an independent nonprofit that provides its product and services for free. We use *Signal* for the rest of this chapter.

15.2   HIPAA Considerations

HIPAA is concerned about securing *Protected Health Information* (PHI) from leakage, but at the same time, requires that instant messaging have an audit trail. This requires that all messaging be logged to a centralized server so the log can be reviewed. In addition, HIPAA requires that the vendor be willing to sign a *Business Associate Agreement*[4] (BAA). As a BAA puts the vendor at a potential liability should their service or software be found responsible for leaking protected health information, you will not find free or inexpensive software that meets HIPAA compliance requirements.

---

[2] *https://wire.com/*

[3] *https://www.signal.org*

[4] *https://www.hhs.gov/hipaa/for-professionals/covered-entities/sample-business-associate-agreement-provisions/index.html*

Most readers and students of this work want to leave *no* record of an encrypted conversation. Also, most of our readers have no need of a BAA.

If your instant messaging needs include HIPAA compliance (this requires meeting Joint Commission guidelines), then the rest of this chapter does not apply to you. I recommend you perform an internet search to find and assess the few options available. Then work with an IT expert to implement your HIPAA-compliant program.

15.3   Signal

Signal is a free platform for peer-to-peer (no centralization) and group secure, end-to-end encrypted communications using instant messaging, voice, and video.

15.3.1   Assignment: Install Signal

In this assignment, you create a *Signal* account. This account allows you to make fully secure, encrypted instant messaging, voice calls, and video conferences with friends and business associates.

1. On your iOS mobile device, open a browser window to *https://signal.org.*

2. Tap *Get Signal*. The App Store opens to *Signal-Private Messenger.*

3. Download and Install *Signal* to your mobile device.

4. On your mobile device, open the *Signal* app.

5. Follow the onscreen instructions to complete the registration process.

15.3.2   Assignment: Invite People to Signal

Before you can communicate with someone else using *Signal* they must also have a *Signal* account.

In this assignment, you invite someone to install *Signal* and create an account.

- Prerequisite: Completion of the previous assignment, or a completed installation of Signal on a mobile device.

1. Open *Signal* on your mobile device (invitations do not yet work with Signal Desktop.)

2. Tap your *profile picture* in the top left corner > *Invite Your Friends*.

3. Select to send either a *Message* or *Mail*.

4. A list of all your phone contacts appears. Select the target contact(s), then tap *Done.*

5. A new email message is created with each of your target contacts listed in the Bcc field, with a link to download *Signal* on their phone.

6. Customize the email to your taste, then tap the *Send* button.

7. Once your target contacts have installed *Signal* on their phone, you receive a text from *Signal* they have joined, and their name appears in your *Signal Contacts* list.

15.3.3 Assignment: Secure Instant Message with Signal.

In this assignment, you instant message your new *Signal* friend.

1. Open *Signal* (for this assignment, on your computer.)

2. From the sidebar, select the desired *Contact*.

3. In the main body area of the *Signal* window, at the bottom in the *Send A Message*, enter a text message for your contact, then tap the *Return* key. The message is sent to your contact and received in seconds.

15.3.4 Assignment: Secure Voice or Video Call with Signal

In this assignment, you make a secure, encrypted voice call to one of your Signal friends.

In this assignment, you make a secure, encrypted voice call to a *Signal* friend.

1. Open *Signal.*

2. Select a *Signal* contact to call.

3. In the top right corner of the *Signal* window tap either the *phone* or the *video* icon.

4. Tap the *Start Call* button.

5. On your friends *Signal* device, they hear their device ringing, and an *Incoming Call* message in *Signal.* if they wish to answer, they tap the *Signal Phone* icon.

6. The two of you can now speak in complete privacy (even better than Maxwell Smart's Cone of Silence[5]).

7. To disconnect, either party taps the *Phone* or *Video* icon.

---

[5] *https://en.wikipedia.org/wiki/Cone_of_Silence*

15.4 Voice, Video, and Instant Message Communications Lessons Learned

- ☐ Phone calls, videoconferencing, text messaging, and other communications over landline, cellular, and internet usually are stored by the manufacturer or developer of the device and carriers.

- ☐ *Signal* is an end-to-end encrypted voice, video, and instant message application for Android, iOS, macOS, and Windows.

- ☐ Only download *Signal* from the developer site at *https://signal.org*.

16 Internet Activity

If you go to a coffee shop or at the airport, and you're using open wireless, I would use a VPN service that you could subscribe for 10 bucks a month. Everything is encrypted in an encryption tunnel, so a hacker cannot tamper with your connection.

–Kevin Mitnick[1]

What You Will Learn in This Chapter

- Understand Virtual Private Network (VPN)

- Understand gateway VPN

- Search for a VPN host

- Install and configure VPN

- File share within a Hamachi mesh VPN

- Resolve email conflicts with VPN

- Understand Domain Name System (DNS)

- Test for DNS Leak

What You Will Need in This Chapter

- [Optional] iCloud+ account (paying for more than default 5GB iCloud storage)

- Access to your home or office router and the administrator credentials

---

[1] *https://en.wikipedia.org/wiki/Kevin_Mitnick*

16.1 Virtual Private Network (VPN)

In case you have been sleep-reading through this book, let me repeat my wake-up call: *They are watching you on the Internet. They* may be the automated governmental watchdogs (of your own or another country), government officials (again, of your own or another country), the administrator or owner of your local area network router, bored staff at an Internet Service Provider or broadband provider, a jealous (slightly wackadoodle) ex, high school kids driving by your home or office or sitting on a hill several miles away, marketing groups, or outright professional criminals.

Regardless, your device, data, and privacy are at risk.

Perhaps one of the most important protection steps you can take is to encrypt the entire Internet experience all the way from your computer, through your broadband provider, to a point where your surfing, chat, webcam, email, etc. cannot be tracked or understood. This is accomplished using a technology called *Virtual Private Network[2] (VPN)*.

16.2 Gateway VPN

There are two fundamental flavors of VPN. The most common is called a *gateway VPN*. Mesh VPN is discussed later. Historically, gateway VPN involved the use of a VPN appliance resident at an organization. Telecommuting staff can use the VPN gateway so the Internet acts like a very long Ethernet cable connecting the staff person's computer to the office network. In addition, all data traveling between the user's computer and the gateway is military grade encrypted. The downside to this strategy is that these appliances are expensive (from $600 to several thousand dollars), and they require significant technical experience to properly configure and maintain.

The gateway VPN concept works like this:

---

[2] *https://en.wikipedia.org/wiki/Virtual_private_network*

1. Your device has VPN software installed and configured to connect to a VPN server at the office. This server is connected to your office network.

2. On your device, you open the VPN software and instruct it to connect to the VPN server. This typically requires entering your authentication credentials of username and password.

3. The VPN server authenticates you and begins the connection between itself and your computer. From this point forward, all data traveling between your device and the VPN server are encrypted. Data between the VPN server to the greater internet is not encrypted, unless connecting to a server or service using encryption itself, such as a website using https.

4. Once your data reaches the VPN server, it is forwarded to the appropriate service on your organization's network such as a file server, printer, and mail server.

Although this may sound complex, all a user must do is enter a name, password, and sometimes a key. Everything else is invisible to the user. The only indicator that anything is different is your speed is slower than normal. This is due to the overhead of the encryption/decryption process.

We can use this same strategy to securely surf the Internet by using a VPN service that acts as an intermediary between your device and the Internet.

Using this strategy of a VPN Internet server, all your Internet traffic is military grade encrypted between your computer and the VPN server. It is not possible to decipher your traffic including usernames, passwords, data, or even the type of data coming and going.

One downside is that once the data exits the VPN server, it is readable. However, your data is intermingled with thousands of other user's data, making the process of tweezing out your data a task that only the NSA can accomplish.

Another concern is that some VPN providers maintain user activity logs. This is the law in most countries, so that government agencies can review who is doing what through the VPN. Ideally, you want to work only with a VPN provider operating in a country that does not require logs, and in fact, does not keep logs.

There are thousands of VPN Internet Servers available. Most of them are free. I do not recommend using the free services for two reasons:

- You get what you pay for. Typically, here today, gone tomorrow, unstable, etc.

- You do not know who is listening at the server side of things. Remember, your data is fully encrypted up to the server. But once the data reaches the server on the way to the Internet, it is readable. There must be a high degree of trust for administration of the VPN server.

There are hundreds of legitimate VPN hosts. There are thousands of illegitimate VPN hosts. When determining the best VPN provider for your use, here are key variables to assess:

- **Privacy Jurisdiction.** Must the VPN host comply with US National Security Letters, or other governmental requirements for disclosure of your information? If you care about anybody accessing all your Internet data, the geographical location of the host headquarters is important.

- **Logs**. Are logs kept on client activities? In many countries, it is required by law that all Internet providers maintain logs of client activities. If so, although the logs may not record *what* you were doing, they keep a record of *where* you traveled. It is ideal to have a VPN provider that keeps no logs.

- **Speed**. How fast is your Internet experience? Using VPN introduces a speed penalty due to the encryption/decryption process, as well as the need to process all incoming and outgoing packets through a server instead of point-to-point. VPN providers can reduce this penalty in several ways: faster servers, reduced clients:server ratio, better algorithms, content filtering to remove advertisements and cookies, and faster server internet connections.

- **Support**. VPN adds a layer of complexity to your Internet activities. Should something not work correctly, you do not want to be the one troubleshooting. Ideally, your VPN provider has 24/7/365 chat support. Even better if they offer telephone support.

- **Cross-Platform Support**. Most of us have more than one device. It would be madness to have to use a different VPN product for each of these. Look for a provider that supports all your current and potential devices.

- **Multi-Device Support**. Most, but not all, providers now offer from 3-5 concurrent devices licensing. This allows your VPN service to be operational on all your devices at the same time. Providers that offer only single-device licensing may be quite costly should you have multiple devices.

- **DNS-Leak Protection**. A VPN encrypts all data that comes and goes from your device. But before you reach out to the Internet to connect to your email, a website, or text, your device must connect to a DNS server for guidance on where to find the mail, web, or text server. If you use your default DNS server (typically one from your Internet broadband provider), the data between your system and the DNS server is not encrypted *and* is recorded. It is ideal if your VPN provider offers their own DNS servers. Using this strategy, the data between your device and the DNS server is now encrypted or not logged.

- **VPN Protocol.** There are several network protocols available for encryption. We currently recommend IKEv2. This is the most current, and one of the few protocols with the ability to automatically activate upon accessing the Internet and deactivate when not in use.

- **Pricing.** This is sometimes directly related to the quality of service, and occasionally directly related to the greediness of those running the business. Look for reasonable pricing for the services offered, as well as how many concurrent connections you are allowed. Some hosts allow only 1 connection. Others offer 6 or more, which allows for you to have your computer, phone, tablet connected via VPN, as well as those of a family member.

Now you may be asking yourself: If VPN is so great, why doesn't everyone know about and use it?

Great question! As with everything else in life, there is bad that comes with the good. Each person needs to weigh the pros and cons for their situation. I *always* have VPN active, but I'm *always* doing work! There are three primary downsides to VPN.

- It slows your Internet performance, often by 50% or more. If all I want to do is to stream Netflix to my computer, I'd turn VPN off to reduce the pauses induced by a slow Internet connection.

- If you have selected a VPN server outside of your home country, you may have unintended consequences due to the *Internet* servers thinking you are resident in that other country. For example, Google searches display in the language native to that country. This is considered a feature of the *Proxy Server* function built into VPN and is used in restrictive countries to view news across the border that normally is filtered out by a home country.

- Some websites and services refuse to work with VPN active. This is usually because they want to be able to track you. You will need to decide which you want more – a particular application or privacy.

16.2.1   Assignment: Search for VPN Host

In this assignment, you search for at least three VPN hosts that meet your needs.

1. Based on the list of criteria listed above, make a list of VPN Host *must-haves* and *prefer-to-haves*.

2. Open a web browser to https://www.safetydetectives.com.

3. From the site menu, select *Best VPN*.

4. From the sidebar, scroll down then select *VPN Comparison Charts*. The URL is https://www.safetydetectives.com/best-vpns/#comparison.

5. There are two charts–*Simple VPN Comparison Chart* and *Detailed VPN Comparison Chart*. For this assignment, select *Simple VPN Comparison Chart*.

6. Tap the *Show* <pop-up menu> *entries*, then select *All*.

7. To sort by your first must-have, tap in that column header. For this assignment, tap on *Privacy Logging*.

8. From the few hosts meeting these minimum requirements, view their websites to evaluate if any meet your needs.

9. Outside this course, you may sign up with one or more of the hosts meeting your needs, put its service through an evaluation, unsubscribe and request a refund if they do not meet with your expectations.

10. Below this *Simple* form is a *Detailed VPN Comparison Chart.*

16.3 NordVPN

One of our favorite VPN providers is *NordVPN*[3]. They offer subscriptions as short as a month, as long as 2 years. With this you get servers in many countries, use on multiple devices, unlimited and highly responsive bandwidth, specialized VPN servers, and a good price.

16.3.1 Assignment: Create A NordVPN Account

In this assignment, you create a 3-day free account with *NordVPN.com*, then configure VPN services.

1. In this assignment, you create a NordVPN account. NordVPN has a 30-day guarantee. If within 30 days you do not see value in the service, request a refund.

2. To open a free 3-day trial account, open a browser then visit *https://free.nordvpn.com/.*

3. Tap the *Start Now* button, then tap *Continue to Payment.*

4. Follow the onscreen instructions to create your account.

5. Open the email from NordVPN, then tap the *Activate Now* button.

6. At the *Set password* page, create a password for your NordVPN account, then tap the *Set Password* button.

7. Open the *App Store.app,* then search for and download the *NordVPN* app.

8. Once downloaded, open the *NordVPN* app.

---

[3] *https://nordvpn.com*

9. Enter your NordVPN *account* and *password*.

16.3.2 Assignment: Configure NordVPN

We typically recommend using the IKEv2[4] protocol for VPN, both for its strong encryption and for its automatic activation when accessing the Internet. However, NordVPN has developed its own VPN protocol, NordLynx, which we have found to be just as secure, but even faster than IKEv2.

In this assignment, you configure a VPN connection with NordVPN using the NordLynx protocol.

● Prerequisite: Completion of the previous assignment.

1. Open the *NordVPN* app downloaded in the previous assignment.

2. In the NordVPN home screen, tap the *Profile* icon in bottom right corner > *Settings* icon in the top left corner. *Settings* open.

_____

[4] *https://en.wikipedia.org/wiki/Internet_Key_Exchange*

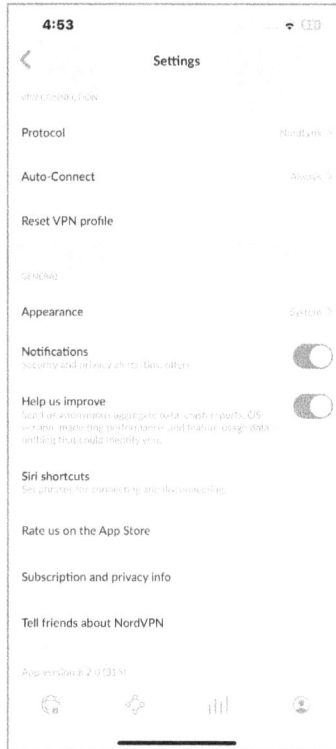

3. Tap *Protocol*. Set Protocol to NordLynx, then tap the *Back* icon.

4. Tap *Auto-Connect*. Set to *Always,* then tap the *Back* icon.

5. Tap the *Back* icon to return to the *Profile* screen.

6. Enable *Threat Protection*. This blocks some unwanted ads and malicious websites.

7. Tap the *Global* icon at bottom left corner to return to the NordVPN Home Screen.

8. Open *Settings > VPN,* then tap the *I* icon.

9. Assuming you have set NordVPN to auto connect, you will see a map with an animated green radar icon indicating the server in use.

Disconnect from VPN

Once you are connected to NordVPN, you can just leave it on. However, should you wish to disconnect:

10. Open the *NordVPN* app.

11. Tap the *Pause* button.

Reconnect To VPN

12. Open the *NordVPN* app.

13. Tap the *Resume* button at bottom center.

16.4   Resolve Email Conflicts with VPN

Some email servers send up a red flag or block user access to email when the user switches to a VPN connection. This is a good thing as it indicates the email provider is highly sensitive to any possible security breach. In all cases, there is a resolution available although the steps to take vary with each provider.

As an example, I list below what occurs when using VPN with a Gmail account, and how to gain access to your email after the blockage.

1. The user starts a VPN program to encrypt all data between the user's computer and the Internet.

2. The user attempts to receive Gmail.

3. Google sees attempted access from an unknown machine (the Proxy Server), and blocks access to the account.

4. Both an email and a text from Google are sent to notify the user of suspicious activity.

5. Select the link in either the email or text message.

6. The first support file opens. Select the link.

7. In the authentication window, enter your email and password, then select the *Sign In* button.

8. Another support window opens, explaining the next steps to take. Select the *Continue* button.

9. The final support window opens. Following the instructions, return to your email application and access your Gmail within 10 minutes. This provides Google with the authentication to release your account.

16.5 Mesh VPN

Another way in which VPN can be configured is a *mesh VPN*. This strategy places multiple computers within the same virtual network regardless of where they are geographically located on the Internet. All the computers operate as if they are on the same physical network, and all traffic between each of the computers is military grade encrypted. Mesh VPN is ideal for groups of people to exchange files, screen share, and access databases from each other, while maintaining full privacy from the outside world.

NordVPN includes *Meshnet* with their service, allowing up to 10 of your own devices and 60 devices from other NordVPN users to share a virtual network at no additional cost.

If you need to include a greater number of devices, or don't want to use NordVPN, *LogmeIn Hamachi*[5] offers virtual networking with almost unlimited devices.

16.6 Domain Name System (DNS)

Most activities on the Internet require pointing to a specific device by use of an address. For example, to use my email, the email software must be able to locate my email server. It does this by looking for *mail.thepracticalparanoid.com*.

Such human-readable names (called a *Fully Qualified Domain Name,* or *FQDN*) work well for you and me, but not so much for computers. Computers expect to use a TCP-IP address. In the case of this server, that is *172.217.3.39*.

---

[5] *https://www.vpn.net*

The translation from FQDN to TCP-IP address is done by way of the *Domain Name System (DNS)*. The process works like this:

1. The user, software, or setting enters the FQDN. For this example, I may enter it in a web browser so that I can view my email.

2. The browser has no idea how to find the FQDN, so it sends the request to the designated DNS server. For macOS, this is configured in *System Settings > Network > Advanced > DNS*.

   o For iOS using Wi-Fi, this is configured in *Settings > Wi-Fi > current Wi-Fi SSID Info* icon *> Configure DNS*.

   o For iOS using cellular, this is automatically configured upon connection to the cellular network. There is no manual configuration possible.

3. The DNS server maintains a database of all registered FQDN and their TCP-IP address. It sends the search result back to my computer, then the browser takes me to my email.

The system is fast and stable. The concern is that your Internet provider is usually your DNS provider. This allows the provider to monitor and log most of your Internet activity without your consent or knowledge.

If you use VPN, and your VPN provider has DNS Leak Protection, your Internet provider cannot see your DNS queries. But you may not be using VPN all the time.

To protect your Internet activity from being logged by your Internet provider, manually configure your DNS server to be one that ensures your privacy. I recommend the 1.1.1.1 and 1.0.0.1 servers hosted by Cloudflare[6]. They do not monitor activity, nor do they maintain logs.

In addition to their 1.1.1.1 and 1.0.0.1 DNS servers which focus on security, privacy, and speed, Cloudflare has two additional free DNS servers:

- Malware blocking only, 1.1.1.2 and 1.0.0.1

- Malware and Adult Content blocking, 1.1.1.3 and 1.0.0.3

---

[6] *https://www.cloudflare.com/learning/dns/what-is-1.1.1.1/*

16.6.1   DNS Leak Test

If you have a DNS leak, your DNS records (including internet travels) may be visible to others.

In this assignment, you test for a DNS leak.

If you do have a DNS leak, your DNS records (including internet travels) may be visible to others.

In this assignment, you test for a DNS leak.

1. Open a browser to *https://www.dnsleaktest.com.*

2. Select the *Extended Test* button.

3. When the test completes, verify your ISP is not listed, and only your desired DNS host or VPN provider displays. This example uses *Cloudflare* for the DNS host (1.1.1.1 and 1.0.0.1) because it provides free, unmonitored, log-free DNS.

16.6.2   Assignment: Secure DNS Traffic over Wi-Fi

In this assignment, you manually configure your DNS settings to use Cloudflare instead of the default (typically your Internet provider) DNS.

1. Open *Settings* > *Wi-Fi* > current Wi-Fi SSID *Info* icon > *Configure DNS.* Configure as below.

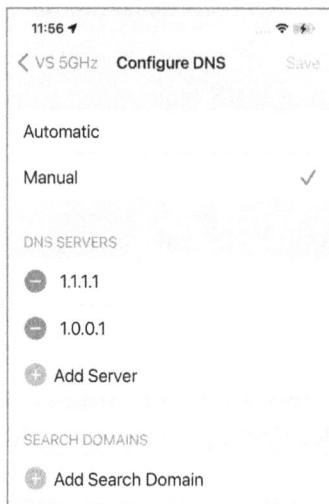

a. Select *Manual.*

b. Delete the existing DNS entries.

c. Tap Add Server.

d. Enter `1.1.1.1` as the first entry.

e. Tap Add Server.

f. Enter `1.0.0.1` as the second entry.

g. Tap *Save.*

h. Exit Settings.

From now on, all your DNS searches are performed securely by Cloudflare (when using Wi-Fi).

16.6.3   Assignment: Secure DNS and Enable VPN with Cloudflare

In this assignment, you configure DNS settings to automatically use Cloudflare instead of the default (typically your Internet provider) DNS over both Ethernet and Wi-Fi. The Cloudflare app provides the option to use the Cloudflare VPN service for free, or to upgrade to higher speeds for fee.

Install Cloudflare 1.1.1.1 app

1. Open *App Store.app,* locate and download *1.1.1.1* from Cloudflare.

2. Open 1.1.1.1.app.

3. At the *What is WARP?* screen, tap *Next.*

4. At the *Install VPN Profile*, tap *Install VPN profile.*

5. At the confirmation prompt to add VPN configurations, tap *Allow.*

6. At the password prompt, enter your login password for your device.

7. At the *Keep up to date* screen, tap *Allow notifications.*

8. At the confirmation prompt, tap *Allow.*

9. At the *WARP* screen, tap the slider to *Connect.*

Verify Your Configuration

10. Open *Settings.app* > *VPN*. The *VPN* screen opens. The *Status* should read *Connected.*

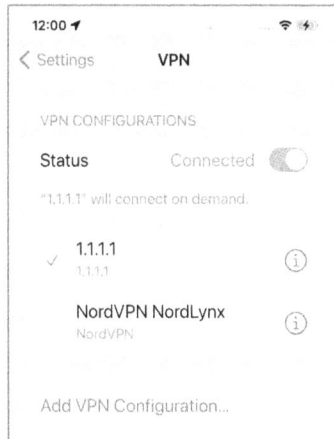

11. Tap the *1.1.1.1* info button. The *Connect On Demand* should be enabled.

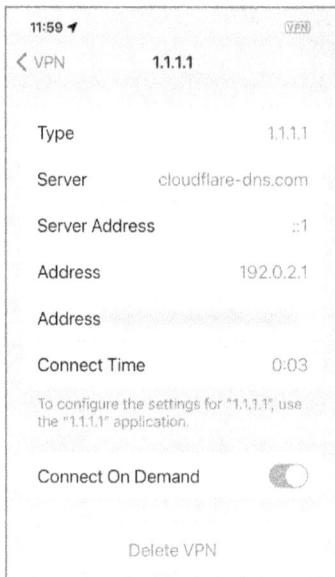

12. When complete, tap < *VPN* then tap < *Settings* to return to the *Settings* screen.

13. If you are currently connected to Wi-Fi, tap *Wi-Fi*, then tap the info button to the right of your current Wi-Fi connection.

14. Scroll down to tap *Configure DNS.* It should be set to *Automatic,* and the *DNS Servers* to include *1.1.1.1* and possibly the address of the Wi-Fi router/gateway currently used.

15. When done, exit *Settings.*

Turn WARP VPN Off

Not all web services or sites behave well when connecting via VPN. If you must turn off your WARP VPN, or prefer to use a different VPN, but keep your secure DNS working, this is your assignment.

1. Open the *1.1.1.1* app. The app opens to the *WARP* screen.

2. Tap the slider to *Disconnect WARP*.

3. From the confirmation screen, tap to select how long you wish WARP to be disabled.

4. Exit 1.1.1.1.

From this point forward, all your cellular and Wi-Fi data and DNS traffic is encrypted.

16.7 Private Relay

Private Relay[7], new with macOS 12, iOS 15, and iPadOS 15, prevents third-party sites from determining your web-browsing habits, and blocks the ability to fingerprint you. The third-party sees an IP address that is not yours and what that IP address is doing on the site. The same is true for anyone else with access to your internet traffic.

This is done by:

• Encrypting all internet-bound traffic when using Safari

• Routing internet-bound traffic from Safari through two proxy servers–one owned by Apple, the other by another provider–which strip off your IP address and separates your IP from the web activity.

As of this writing, Private Relay is available to all iCloud+ users (paid iCloud) in most but not all countries. Private Relay will apply to:

• All Safari web browsing

---

[7] *https://www.apple.com/newsroom/2021/06/apple-advances-its-privacy-leadership-with-ios-15-ipados-15-macos-monterey-and-watchos-8/*

- All DNS queries as users enter site names
- All insecure HTTP traffic

Private Relay will not apply to:

- Local Area Network
- Private domain name queries
- Traffic using regular VPN
- Internet traffic using a proxy

Private Relay is different from regular VPN:

Feature	Private Relay	VPN
Hide true IP address	☑	☑
Unblock websites blocked by ISP	☑	☑
Works with all browsers and mail apps Apple states Private Relay only works with Safari. However, as of this writing Private Relay works with all browsers and the Mail.app.	⊘?	☑
Unblock geo restricted content Private Relay gives no option for user to select apparent geo location, but it is always within the same country. VPN allows choosing from many countries.	⊘	☑
Cost Private Relay is free (after paying for iCloud+). Free VPN should never be used.	☑	⊘
Security Private Relay uses a dual-hop architecture, not even Apple can decipher the data. VPN can see your activity locations.	☑	☑?

16.7.1 Assignment: Use Private Relay

In this assignment, you configure Private Relay to automatically activate when using Safari.

- Prerequisite: iOS 15 or higher

- Prerequisite: Safari

- Prerequisite: iCloud+ account. If you are paying for more than the default 5GB of iCloud Drive storage, you have an iCloud+ account.

Verify iCloud+ In Use

By default, all iOS users have an iCloud account, tied to their Apple ID, with 5GB of cloud storage. If you see your account has more than 5 GB of cloud storage, you have an iCloud+ account. If you do not have an iCloud+ account, you can upgrade from the following screen.

1. Open *Settings > Apple ID.*

2. Tap your name. This opens the *Apple ID* screen.

3. Scroll down to the *iCloud* field. If you have 5 GB of storage space, you may either skip this assignment, or continue to upgrade to iCloud+. If you have more than 5 GB of storage space, skip to step ___ of this assignment.

4. Tap *iCloud > Manage Storage > Change Storage Plan,*

5. In the *Upgrade iCloud Storage* screen, select your desired plan:

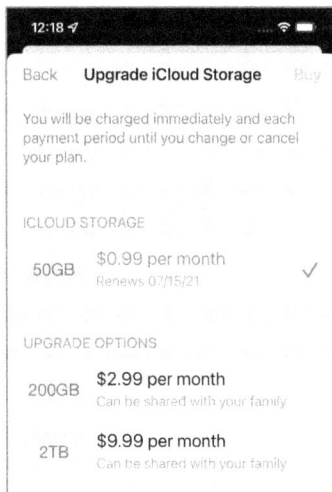

6. Follow the onscreen instructions to complete the subscription process.

Enable Private Relay

7. Return to the *iCloud* screen (*Settings* > tap your name > *iCloud*).

8. Tap *Private Relay*. The *Private Relay* screen opens:

9. Tap *IP Address Location*. Select your desired location option, then tap *Back*.

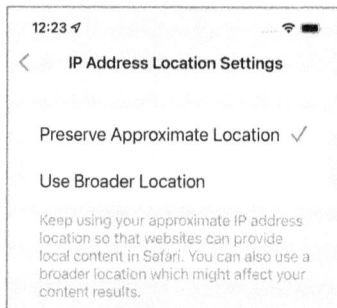

10. In the *Private Relay* screen, turn on *Private Relay*.

11. Exit *Settings*.

Test Private Relay

12. Verify you do not have any VPN active, including Cloudflare 1.1.1.1.

13. Open Safari to *https://whatismyip.com*.

14. Leaving Safari open, open another browser to *https://whatismyip.com*.

15. View the results side-by-side. For this assignment, I am physically located in Santa Fe, New Mexico, USA. The first image is from Safari, the second is from Brave:

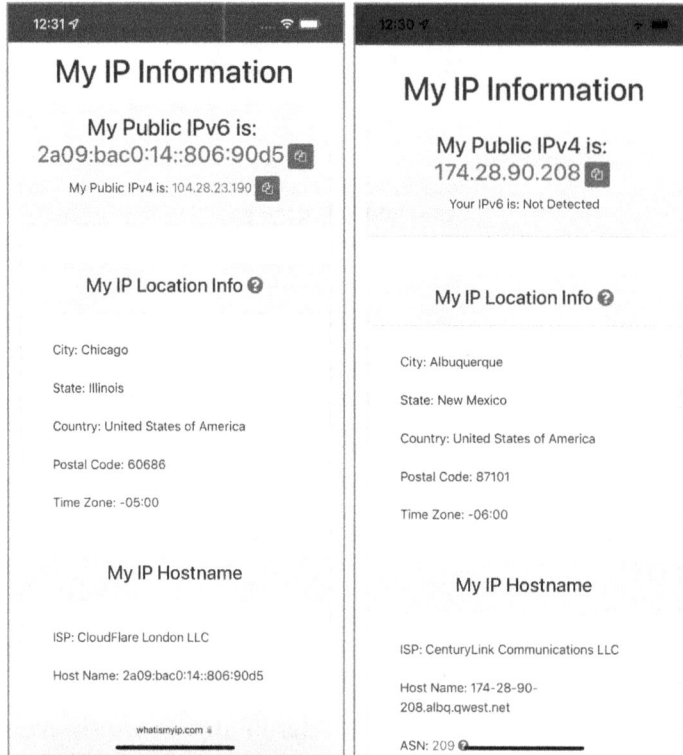

16.8 Cybersecurity and Privacy Lessons Learned

☐ Virtual Private Network (VPN) encrypts data between your device and the VPN server.

☐ Gateway VPN allows devices to connect with the internet using encryption between the device and the VPN server.

☐ Mesh VPN allows people and devices connected to the internet while geographically separated to function as though all devices are on the same encrypted local network.

- [ ] Even if using VPN, your internet provider or VPN provider may be able to track your activities by your DNS activity. This is called a DNS leak. The better VPN providers have built-in secure encrypted DNS services.

- [ ] By enabling the Cloudflare 1.1.1.1 app for DNS, you always have secure DNS with both cellular and Wi-Fi connections.

- [ ] Some websites and services refuse to work with VPN active. This is usually because they want to be able to track you. You will need to decide which you want more – a particular application or privacy.

- [ ] Private Relay is an Apple-only feature, available to iCloud+ account holders, using macOS 12, iOS 15, and iPadOS 15, with Safari. It encrypts all internet-bound traffic and separates your IP address from your web activities.

17 Social Media, Apple ID and iCloud

A lot of people say that social media is making us all dumber, but I not think that.
–Unknown author

What You Will Learn in This Chapter

- Protect your privacy on social media
- Harden Facebook security
- Harden LinkedIn security
- Harden Google security
- Harden Apple ID & iCloud security

What You Will Need in This Chapter

- No additional resources required.

17.1   What, Me Worry?[1]

Social media[2] has provided us with new levels of connectivity and communication. Victims of natural and man-made catastrophes can instantly assure family and friends of their location and health. Job searches have been reduced to a few mouse taps. And those who once would have had no voice, may now have a voice that is heard around the world.

---

[1] *Alfred E. Neuman, https://en.wikipedia.org/wiki/Alfred_E._Neuman*
[2] *https://en.wikipedia.org/wiki/Social_media*

Virtually all social media is free to the user. Few users ever question how a service such as Facebook, that may have operating expenses of 12 *billion* dollars[3], not only can afford such expense, but then go on to have a profit of 10 *billion* dollars. The business model for social media is partially based on advertising, but far more on selling information about *you* to the advertisers.

Social media knows more about you than your mother. Their systems know what you are doing on their site–as well as all ancillary sites–including how long you stay on each page, where you came from to land on that page, where you jump off to, what you have purchased, your interests, issues that prompt strong emotion in you, and more. The data and metadata held by social media sites have been shown to be extremely accurate predictors of behavior. It is *this* information that is so very valuable to advertisers, and potential employers, and the government.

It has become the norm for HR departments to scan all social media of a potential employee. The belief is that what a person expresses in social media is a more accurate and honest representation than the employment form or initial interview.

Government agencies closely track social media to predict and stop the next terrorist attack, as well as other lesser crimes.

Then we have criminals tracking social media for mentions of things like "*...our entire family is leaving for a month-long vacation next Friday....*"

17.2   Protect Your Privacy on Social Media

Privacy on social media starts with one step: understand that social media watches *everything* you do, when you do it, how long you do it, and who you do it with. Social media also becomes your *brand*, intentional or not. It colors how others– friends, family, employers, and government–see you. To this effect, you must manage your brand. This is done by being fully mindful of *everything* about your social media pages. Does it represent you in the best light? If taken out of context, how would your data be interpreted?

---

[3] *https://finance.yahoo.com/quote/FB/financials?ltr=1*

The next step is to take whatever measures the social media site offers you to ensure that only those you *want* in, do get in. In the following sections, we show how to protect your security using Facebook, LinkedIn, Google, Apple ID and iCloud privacy features.

17.3 Facebook

Facebook is the reigning king of social media. Whether due to lax security concerns, or the wild popularity of the site, account breaches and hacks are common. Fortunately, Facebook has taken user-centric security controls out of the shadows and made them easily accessible.

Facebook collects information about you from three different levels:

- Data about your interactions while on Facebook.
- Data about your interactions with websites that have Facebook tracking.
- According to a *Ghostery*[4] report, Facebook has trackers on 23% of websites.
- Data about your interactions with apps that have Facebook tracking.
- As reported in *Slashdot*[5], 61 of the top 100 apps have Facebook tracking.

17.3.1 Assignment: Facebook Security and Login

In this assignment, you secure your Facebook password.

• Prerequisite: Ownership of a Facebook account.

1. Open a browser to *https://facebook.com/*, then log into your account.
2. Tap the downward triangle at the top right corner of the Facebook window. Select *Settings & Privacy > Settings*.

---

[4] *https://www.ghostery.com/tracking-the-trackers-2020-web-trackings-opaque-business-model-of-selling-users/*

[5] *https://tech.slashdot.org/story/21/08/29/1758218/facebook-has-trackers-in-25-of-websites-and-61-of-the-most-popular-apps*

3. From the sidebar, select *Security and Login.* Carefully review and configure each of the areas and settings. Of particular importance:

   a. Are any of the *Where You're Logged In* devices not yours?

   b. Change your password to a strong, unique password.

   c. Enable Two-Factor Authentication (2FA).

4. From the sidebar, select *Privacy.* Carefully review and configure each of the areas and settings.

5. From the sidebar, select *Face Recognition.* This should be disabled for most of us. If enabled, Facebook can use this to track all images with you in them.

6. From the sidebar, select *Profile and Tagging.* Carefully review and configure each of the areas and settings.

7. From the sidebar, select *Public Posts.* Carefully review and configure each of the areas and settings.

8. From the sidebar, select *Blocking.* Carefully review and configure each of the areas and settings.

9. From the sidebar, select *Location.* This should be disabled for most of us.

10. From the sidebar, select *Stories.* This should be disabled for most of us.

11. From the sidebar, select *Apps and Websites.* Carefully review and configure each of the areas and settings.

12. From the sidebar, select *Instant Games.* Games are a huge security vulnerability. I recommend removing any games listed.

13. From the sidebar, select *Business Integration.* Unless you trust and value the service offered here, remove them.

14. From the sidebar, select *Ads > Ad Settings.* Carefully review and configure each area and setting, drilling down the multiple levels you find. This section is critically important to help restrict your Facebook activities from being known by marketeers.

17.3.2 Assignment: What Does Facebook Know About You

Much of the information Facebook has about you may be found in the assignments above. But there is more that can be mined. In this assignment, you have Facebook provide the information it holds and shares about you.

The data that Facebook has on each member includes the following: (For a full list, visit *https://www.facebook.com/help/405183566203254*).

What info is available?	What is it?	Where can I find it?
About Me	Information you added to the About section of your Timeline like relationships, work, education, where you live and more. It includes any updates or changes you made in the past and what is currently in the About section of your Timeline.	Activity Log Downloaded Info
Account Status History	The dates when your account was reactivated, deactivated, disabled, or deleted.	Downloaded Info
Active Sessions	All stored active sessions, including date, time, device, IP address, machine cookie and browser information.	Downloaded Info
Ads Clicked	Dates, times, and titles of ads clicked (limited retention period).	Downloaded Info
Address	Your current address or any past addresses you had on your account.	Downloaded Info
Ad Topics	A list of topics on which you may be targeted based on your stated likes, interests, and other data you put in your Timeline.	Downloaded Info
Alternate Name	Any alternate names you have on your account such as a maiden name or a nickname.	Downloaded Info
Apps	All the apps you have added.	Downloaded Info
Birthday Visibility	How your birthday appears on your Timeline.	Downloaded Info
Chat	A history of the conversations you've had on Facebook Chat (a complete history is available directly from your messages inbox).	Downloaded Info
Check-ins	The places you've checked into.	Activity Log Downloaded Info

What info is available?	What is it?	Where can I find it?
Connections	The people who have liked your Page or Place, RSVPed to your event, installed your app or checked in to your advertised place within 24 hours of viewing or clicking on an ad or Sponsored Story.	Activity Log
Credit Cards	If you make purchases on Facebook (for example: in apps) and have given Facebook your credit card number.	Account Settings
Currency	Your preferred currency on Facebook. If you use Facebook Payments, this will be used to display prices and charge your credit cards.	Downloaded Info
Current City	The city you added to the About section of your Timeline.	Downloaded Info
Date of Birth	The date you added to Birthday in the About section of your Timeline.	Downloaded Info
Deleted Friends	People you've removed as friends.	Downloaded Info
Education	Any information you added to Education field in the About section of your Timeline.	Downloaded Info
Emails	Email addresses added to your account (even those you may have removed).	Downloaded Info
Events	Events you've joined or been invited to.	Activity Log Downloaded Info
Facial Recognition Data	A unique number based on a comparison of the photos you're tagged in. This data helps others tag you in photos.	Downloaded Info
Family	Friends you've indicated are family members.	Downloaded Info
Favorite Quotes	Information you've added to the Favorite Quotes section of the About section of your Timeline.	Downloaded Info
Followers	A list of people who follow you.	Downloaded Info
Following	A list of people you follow.	Activity Log
Friend Requests	Pending sent and received friend requests.	Downloaded Info
Friends	A list of your friends.	Downloaded Info

What info is available?	What is it?	Where can I find it?
Gender	The gender you added to the About section of your Timeline.	Downloaded Info
Groups	A list of groups you belong to on Facebook.	Downloaded Info
Hidden from News Feed	Any friends, apps, or pages you've hidden from your News Feed.	Downloaded Info
Hometown	The place you added to hometown in the About section of your Timeline.	Downloaded Info
IP Addresses	A list of IP addresses where you've logged into your Facebook account (does not include all historical IP addresses as they are deleted according to a retention schedule).	Downloaded Info
Last Location	The last location associated with an update.	Activity Log
Likes on Others' Posts	Posts, photos, or other content you've liked.	Activity Log
Likes on Your Posts from others	Likes received on your own posts, photos, or other content.	Activity Log
Likes on Other Sites	Likes you've made on sites off Facebook.	Activity Log
Linked Accounts	A list of the accounts you've linked to your Facebook account	Account Settings
Locale	The language you've selected to use for Facebook.	Downloaded Info
Logins	IP address, date and time associated with logins to your Facebook account.	Downloaded Info
Logouts	IP address, date and time associated with logouts from your Facebook account.	Downloaded Info
Messages	Messages you've sent and received on Facebook. Note that if you've deleted a message, it will not be included in your download as it has been deleted from your account.	Downloaded Info
Name	The name on your Facebook account.	Downloaded Info
Name Changes	Any changes you've made to the original name you used when you signed up for Facebook.	Downloaded Info

What info is available?	What is it?	Where can I find it?
Networks	Networks (affiliations with schools or workplaces) that you belong to on Facebook.	Downloaded Info
Notes	Any notes you've written and published to your account.	Activity Log
Notification Settings	A list of all your notification preferences and whether you have email and text enabled or disabled for each.	Downloaded Info
Pages You Admin	A list of pages for which you are the admin.	Downloaded Info
Pending Friend Requests	Pending sent and received friend requests.	Downloaded Info
Phone Numbers	Mobile phone numbers you've added to your account, including verified mobile numbers you've added for security purposes.	Downloaded Info
Photos	Photos you've uploaded to your account.	Downloaded Info
Photos Metadata	Any metadata that is transmitted with your uploaded photos.	Downloaded Info
Physical Tokens	Badges you've added to your account.	Downloaded Info
Pokes	A list of who has poked you and who you have poked. Poke content from our mobile poke app is not included because it's only available for a brief time. After the recipient has viewed the content, it's permanently deleted from your system.	Downloaded Info
Political Views	Any information you added to Political Views in the About section of Timeline.	Downloaded Info
Posts by You	Anything you posted to your own Timeline, like photos, videos, and status updates.	Activity Log
Posts by Others	Anything posted to your Timeline by someone else, such as wall posts or links shared on your Timeline by friends.	Activity Log Downloaded Info
Posts to Others	Anything you posted to someone else's Timeline, such as photos, videos, and status updates.	Activity Log
Privacy Settings	Your privacy settings.	Privacy Settings Downloaded Info

What info is available?	What is it?	Where can I find it?
Recent Activities	Actions you've taken and interactions you've recently had.	Activity Log Downloaded Info
Registration Date	The date you joined Facebook.	Activity Log Downloaded Info
Religious Views	The current information you added to Religious Views in the About section of your Timeline.	Downloaded Info
Removed Friends	People you've removed as friends.	Activity Log Downloaded Info
Screen Names	The screen names you've added to your account, and the service they're associated with. You can also see if they're hidden or visible on your account.	Downloaded Info
Searches	Searches you've made on Facebook.	Activity Log
Shares	Content (ex: a news article) you've shared with others on Facebook using the Share button or link.	Activity Log
Spoken Languages	The languages you added to Spoken Languages in the About section of your Timeline.	Downloaded Info
Status Updates	Any status updates you've posted.	Activity Log Downloaded Info
Work	Any current information you've added to Work in the About section of your Timeline.	Downloaded Info
Vanity URL	Your Facebook URL (ex: username or vanity for your account).	Visible in your Timeline URL
Videos	Videos you've posted to your Timeline.	Activity Log Downloaded Info

1. Open a web browser, then log into your Facebook account.

2. In the top right corner, tap the *triangle* icon > *Settings* > *Your Facebook Information* in the sidebar.

3. Select *Download a Copy of Your Information.*

4. In the *Download Your Information* page, in the *Date Range* field, select *All of my data,* then tap *Create File.*

- It may take several hours for your file to be ready. Facebook sends you an email notifying you when it is ready, along with a download link.

5. In the *Download Your Information* page, select the *Start My Archive* button.

6. At the authentication prompt, enter your Facebook password, then select the *Submit* button.

7. Select the *Start My Archive* button.

8. Your request is submitted to Facebook. Depending on the size of the database, it may take a day or more for you to receive an email link to download your data. Select the *OK* button.

9. Watch for your Facebook link in your email. When the link arrives, tap the link to access your archive.

Manage your Activity Log

10. From the sidebar, select *Your Facebook Information > Activity Log.*

11. The sidebar displays all your recent history.

12. To delete records of specific activities, tap on the activity, tap on the ... button, then select *Delete.*

Manage your off-Facebook Activity Log

13. From the sidebar, select *Your Facebook Information > Off-Facebook Activity.*

14. Select *Manage Your off-Facebook Activity.* A list displays of recent activity that other businesses and organizations shared with Facebook.

15. If you do not want an organization you met on Facebook to share information that organization has gathered about you with Facebook, notify the organization to not share your information by tapping: *Turn off future activity from <company name>.*

16. Exit from Facebook *Settings.*

17.3.3   Assignment: What Does Off-Facebook Know About You

Websites and apps that have Facebook tracking installed will feed your activities on the sites and apps to Facebook. You can view which sites and apps are involved, and the ability to manage this information. More detailed information may be found here.[6]

In this assignment, you view your off-Facebook tracking.

View Off-Facebook Activity

3. Open a browser then log into your Facebook account.

4. Select your *Profile Picture > Settings & Privacy > Settings*.

5. In the *Settings* page, select *Your Facebook Information* from the sidebar > *Off-Facebook activity > View*.

6. The *off-Facebook activity* page opens.

7. Select *Recent Activity* to view where your Facebook data has been.

8. For this assignment, close the *Activity Details* window.

17.4 LinkedIn

While Facebook is the current reigning king of non-business social media, LinkedIn holds the crown for business. Whether it be to market one's services, look for a new job, or simply network with other businesspeople, LinkedIn is the place to be.

Just as with all other social media sites, it is vital to be mindful of privacy and security on LinkedIn. It could be business suicide to have anything but the very best information associated with your account.

17.4.1 Assignment: LinkedIn Account Security

The need for a strong LinkedIn password is no different than for your computer.

In this assignment, you change your current LinkedIn password to a strong password.

---

[6] *https://www.facebook.com/help/2207256696182627*

- Prerequisite: Access to a LinkedIn account.

1. Open a browser to *https://linkedin.com/.*

2. From the tool bar, select the *Me* icon > *Settings & Privacy.*

3. The *How LinkedIn uses your data* page opens.

4. From the sidebar, select *Account Preferences.* The *Account Preferences > Profile Information* page opens.

5. One by one, select each section starting with *Name, location, and industry,* to verify the information is accurate, places you in the best light, and is shared only with those who should see this information.

6. Repeat steps 4 and 5 for each sidebar section.

17.4.2   Assignment: Find What LinkedIn Knows About You

In this assignment, you download all the data that LinkedIn (admits) to knowing about you.

1. Open a web browser, then log into your LinkedIn account.

2. From the *Me* menu in the top right of the page, select *Settings & Privacy.*

3. From the sidebar, select *Data Privacy.* The *How LinkedIn uses your data* page opens.

4. Scroll down to tap *Get a copy of your data.*

5. In the *Get a copy of your data* area, select *Download larger data archive*, then tap *Request archive* button. You are notified by email when the report is available for download.

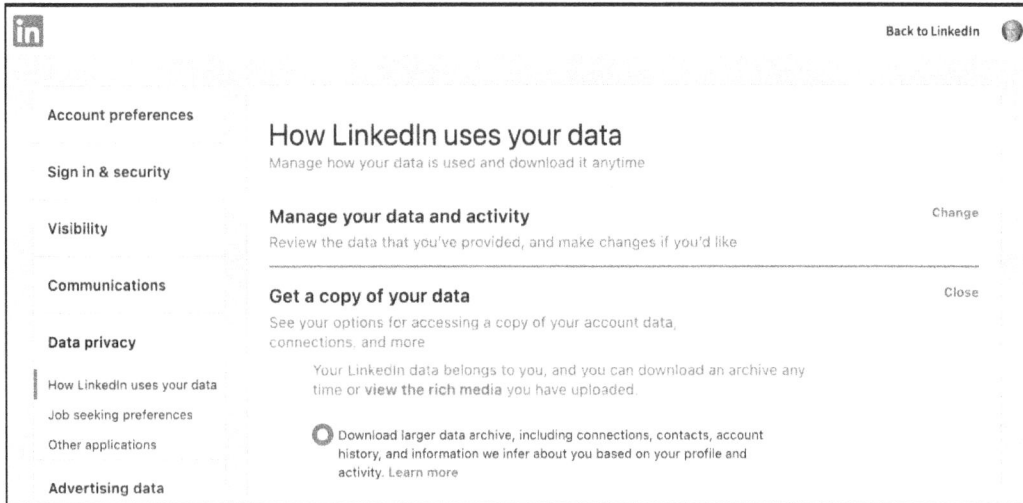

17.5   Google

Although most people think of Google as a search engine, it has become far more than that. Over one billion people use Google mail, maps, YouTube, and Google Play services[7]. Chances are you use Google every day. The result is a tremendous warehouse of data points about your searches, site visits, hangout partners, purchases, and so much more.

And yet, at the same time Google provides the tools to help guard your privacy. NOW is a good time to put these tools to use.

17.5.1   Assignment: Google Sign-in & Security

In this assignment, you begin the process of securing a Google account.

● Prerequisite: Access to an existing Google account.

1. Open a web browser to sign in at the Google Security page *https://myaccount.google.com*.

---

[7] *https://www.digitaltrends.com/web/gmail-joins-the-billion-users-club/*

Personal Info

2. From the sidebar, select *Personal Info.*

3. In the *Basic Info* area, verify all information is accurate. This is also a good time to change your Google password to be more secure.

4. In the *Contact info* area, verify all information is accurate and complete.

5. In the *Choose what others see* area, tap *Go to About Me.*

6. In the *About Me* page, verify all information is accurate and complete, and is viewable only by those who should see the information.

7. When complete, go *Back.*

Data & Personalization

8. From the sidebar, select *Data & Privacy.*

9. Scroll to *Privacy suggestions available > Review suggestions.* Follow the on-screen instructions to improve your privacy, then go *Back.*

10. Scroll down to *History Settings.* Configure *Web & App Activity, Location History,* and *YouTube History* to your taste. I recommend *Pausing* each, to prevent Google building a history on your activity.

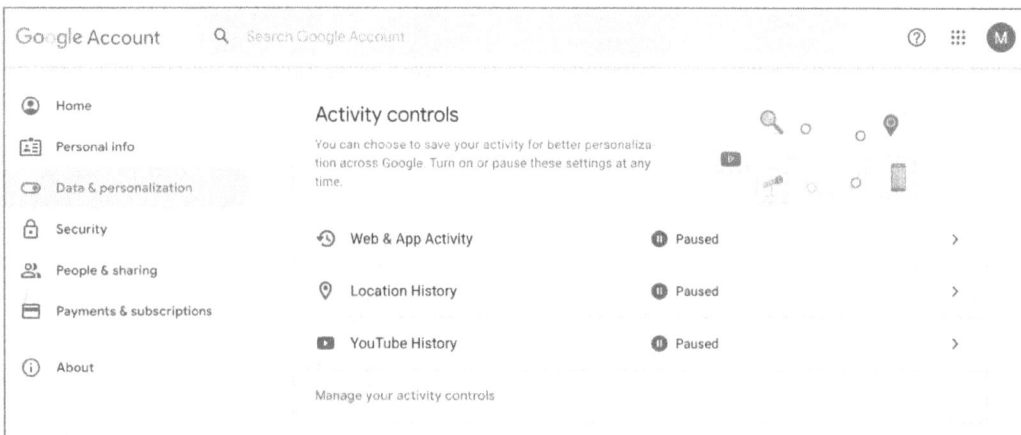

11. In the *Personalized ads* area, turn *off* the following: *My Ad Center,*

12. In the *Info you can share with others* area, configure to your taste, remembering that the less you share, the more secure your data.

13. Tap *Delete a Google service,* authenticate, then review all Google services in use. If a service is active that you no longer use, *delete* it, then go *Back.*

Security

14. From the sidebar, select *Security.*

15. In the *You have security recommendations,* tap *Protect your account.* Review the recommendations, configure to your taste. Then go *Back.*

16. In the *Signing into Google* area, verify you have *2-Stepp Verification* active.

17. In the *Ways we can verify it's you.* Verify your *Recovery phone* and *Recovery email* are accurate and current.

18. Scroll down to *Your devices.*

19. If any of the displayed devices are no longer in use by you, tap *Manage devices, Sign Out* of the device. Then go *Back.*

20. In the *Third-party apps with account access*, if you no longer use any of these apps, tap *Manage third-party access,* select the target app, tap *Remove Access.* Then go *Back.*

21. Scroll down to *Signing into other sites.*

22. In the *Signing in to other sites* area, tap *Signing in with Google.* if you find any apps or services using your Google account to sign in (which means they have access to your Google activities), consider removing them to force them to log in directly. Then go *Back.*

23. Select *Password Manager.* If there are any passwords stored in Google that you do not want there, remove them. Then go *Back.*

24. Select *Linked Accounts.* If you have any linked accounts which give Google access to your data, you may manage that linkage here. Then go *Back.*

People & Sharing

25. From the sidebar, select *People & sharing > Choose what others see > About me.* Configure to your taste, being aware of what information is being made to the world. Then go *Back.*

26. Select *Current settings.* Configure to your taste, being aware of what information is being made to the world. Then go *Back.*

27. Select *Location sharing > Manage location sharing.* Verify you are not sharing your location. Then go *Back.*

17.5.2 Assignment: Find What Google Knows About You with Takeout

In this assignment, you discover what Google knows about you. Google provides a service known as *Takeout* that allows you to download anything and everything Google knows about you (what they admit to knowing about you). When using Takeout, Google delivers to you this data in .zip files. If your history with Google is extensive, the data files are large.

1. Open a web browser, then visit Google Takeout at *https://takeout.google.com/.*

2. From the list of available data subjects, select what you wish to access. I recommend selecting everything. Doing so not only gives you knowledge of what Google knows about you but also provides a backup of all your Google Account data.

← Google Takeout

Your account, your data.
Export a copy of content in your Google Account to back it up
or use it with a service outside of Google.

CREATE A NEW EXPORT

1   Select data to include

Products

Deselect all

Android Device Configuration Service
Android device attributes, performance data, software versions, and account
identifiers. More info

HTML format

Arts & Culture
Favorites and galleries you've created on Google Arts & Culture.

Multiple formats

Calendar
Your calendar data in iCalendar format. More info

iCalendar format   All calendars included

Chrome
Bookmarks, history, and other settings from Chrome More info

Multiple formats   All Chrome data included

Classic Sites
The content and attachments of your sites created in Classic Sites. More info

Multiple formats

Classroom
Your Classroom classes, posts, submissions, and rosters More info

JSON format

Contacts
Contacts and contact photos you added yourself, as well as contacts saved
from your interactions in Google products like Gmail. More info

vCard format

Crisis User Reports
Information provided to help others during crises

Currents Circles
Your Currents circles. More info ☑

📄 vCard format

Currents Communities
Currents communities you own or moderate. More info ☑

📄 Multiple formats

Currents Stream
Your Currents posts, collections, and other contents. More info ☑

📄 Multiple formats　　🔲 All Currents Stream data included

Data Shared for Research
Responses saved with your Google Account from your participation in Google ☑
research studies and projects.

📄 Multiple formats

Drive
Files you own that have been stored in your My Drive and Computers. More ☑
info

📄 Multiple formats　　⚙ Advanced settings　　🔲 All Drive data included

Fit
Your Google Fit activity data. ☑

📄 Multiple formats　　🔲 All Fit data included

Google Account
Data about registration and account activity ☑

📄 HTML format

Google Cloud Search
Content metadata ingested as part of Google Cloud Search indexing flow, and ☑
user search activity data. More info

📄 Multiple formats

Google Help Communities
Your ask and reply contributions to the Google Help Communities including ☑
text and images posted. More info

📄 Multiple formats

Google My Business
All data related to your business. More info ☑

Google Pay
Your saved passes, activity using virtual account numbers, and transaction history from Google services, like Play and YouTube and peer to peer payments, in the Google Pay app. More info

☑

Multiple formats   All activity and saved items included

Google Photos
Your photos and videos from Google Photos. More info

☑

Multiple formats

Google Play Books
The titles and authors of your purchased and uploaded books in Google Play Books, plus notes and bookmarks More info

☑

Multiple formats

Google Play Games Services
Data, including achievements and scores, from games you play More info

☑

Multiple formats

Google Play Movies & TV
Your Google Play Movies & TV preferences, services, watchlist, and ratings. More info

☑

CSV format   All types included

Google Play Store
Data about your app installs, ratings, and orders

☑

Multiple formats

Google Shopping
Google Shopping order history, loyalty, addresses and reviews

☑

Multiple formats

Google Translator Toolkit
Documents you have in your Google Translator Toolkit

☑

Multiple formats

Google Workspace Marketplace
Metadata which describes an application published in Google Workspace Marketplace.

☑

Multiple formats

Groups
Data for your usage of Google Groups and for Google Groups that you own. More info

☑

Hangouts
Your conversation history and attachments from Hangouts.

Multiple formats

Home App
Device, room, home and history information from the Home App. More info

Multiple formats

Keep
Notes and media attachments stored in Google Keep. More info

Multiple formats

Mail
Messages and attachments in your Gmail account in MBOX format. User settings from your Gmail account in JSON format. More info

Multiple formats   All Mail data included

Maps
Your preferences and personal places in Maps

Multiple formats   All Maps data included

Maps (your places)
Records of your starred places and place reviews. More info

Multiple formats

My Activity
Records of your activity data, along with image and audio attachments. More info

Multiple formats   All activity data included

News
Data about the magazines, categories, and sources you are interested in.

TXT format

Pinpoint
Files you have uploaded to Pinpoint

ZIP format

Posts on Google
Your Posts On Google history data including the collections of account, posts, cameos, metrics data, and all uploaded images and videos on Posts on Google and Cameos. More info

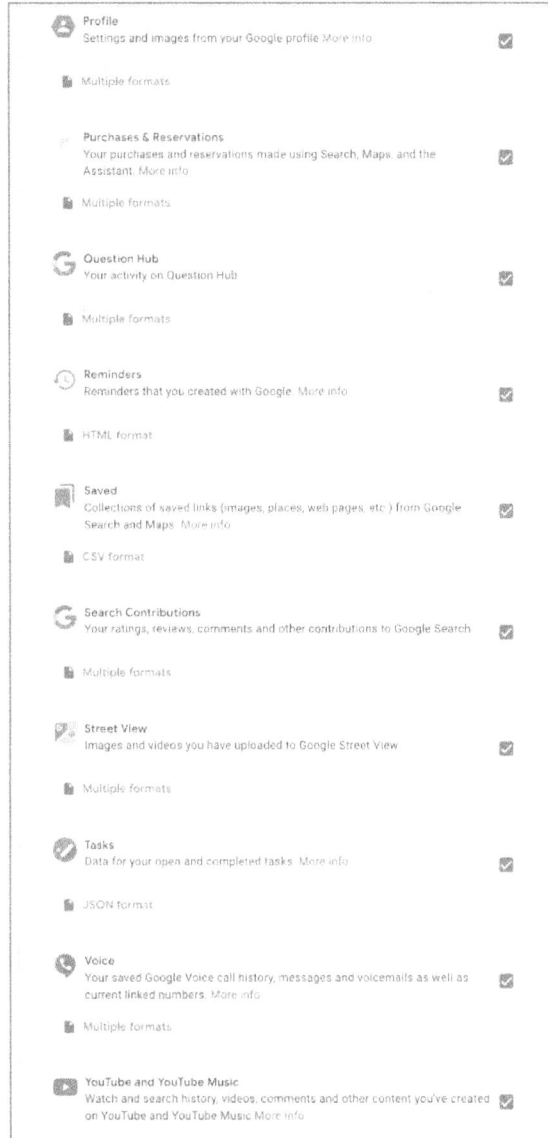

3. Scroll to the bottom, then tap the *Next step* button.

4. In the *Google Takeout Create a New Export* page, configure how you wish to receive your *Takeout*. Then select *Create export*.

5. You receive an email when your archives are available.

6. Once all archives are downloaded, double tap to open each. You may find there are duplicate folders at the root level (such as *Google Drive*). Combine the contents of these duplicates.

7. Have fun learning about yourself!

17.6   Microsoft and Windows Security

In 2013[8] Microsoft's built in Security Essentials app failed certification.

In 2015[9], after multiple patches and official sounding remedies were released, it failed again and truly earned its perennial place, positioned on the bottom of all available anti-virus programs.

In 2016[10], Microsoft inadvertently published a Secure Boot "golden key" policy which allowed for anyone to load self-signed or unsigned binaries onto Windows devices. Rumor had it that these were law enforcement backdoors, built into Windows because Microsoft was not willing to fight the FBI over such things. On the bright side, this event gave Apple an easy win in their fight with the FBI about creating such keys in macOS and iOS.

Skip ahead to 2021[11] and appears that MS has gotten everyone with an outlook exchange account hacked. So, please believe me when I say, you cannot leave Windows 11 security to Microsoft Developers. You must take care of your data, accounts, and access yourself.

Remember that every password and security measure can be broken. Your defense is to make it so difficult and time consuming to break that the hacker

---

[8] *https://www.theverge.com/2013/1/17/3885962/microsoft-security-essentials-fails-anti-virus-certification-test*

[9] *https://redmondmag.com/articles/2015/01/27/security-essentials-fails-antivirus-test.aspx*

[10] *https://www.zdnet.com/article/microsoft-secure-boot-key-debacle-causes-security-panic/*

[11] *https://krebsonsecurity.com/2021/03/at-least-30000-u-s-organizations-newly-hacked-via-holes-in-microsofts-email-software/*

moves on to an easier target. Also, most security questions can be accurately guessed or broken through social engineering (*What is your birthday? In what city did your parents marry? What is the name of your first pet?* etc.) Both types of security are based on what you know. And if there is something that you know, someone else can know it as well.

- Note: As of August 2016, NIST has stopped recommending Two-Factor Verification that involves SMS/text messaging as the second factor[12]. This is due to the ease of which this can be intercepted. However, *any* verification that is received via cellular signal (SMS, voice, etc.) is subject to the same vulnerabilities. Currently, the best solution is to use a digital token (a keychain-sized device that displays random number strings), or an *Authenticator* app, which serves the same purpose.

17.6.1   Assignment: Enable Microsoft 2-Factor Authentication

In this assignment, you enable 2FA for your Microsoft account.

- Prerequisite: Access to a Microsoft account.

- Prerequisite: Access to a smartphone.

- Open a browser to *https://account.microsoft.com.*

- Sign in with your Microsoft credentials.

- Tap your avatar in the upper right corner, then tap *My Microsoft Account.*

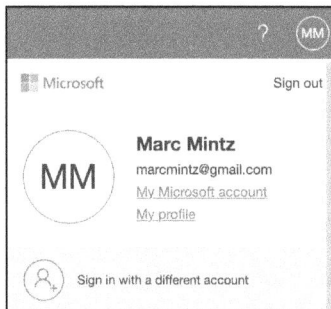

- Tap *Security* in the top menu.

---

[12] *https://pages.nist.gov/800-63-3/sp800-63-3.html*

- Tap *Advanced Security Options.*

- In the *Two-step verification* area, tap to *Turn on.*

- Verify all options are *Up to date.*

- Verify *Two-step verification* is *On.*

- Scroll down to the *Recovery code* area. Make sure to copy your recovery code, print it, and put it with your other account codes somewhere safe.

17.6.2 [Optional] Assignment: Remove a Device from Your Microsoft Account

All the devices (computers, tablets, and phones that you have ever signed into your MS account from, may still have access to your account, presenting a security vulnerability.

To prevent this from happening, you must remove the device from your Microsoft Account device list.

In this assignment, you remove a device from your Microsoft device list.

1. Open a browser to *https://account.microsoft.com/.*

2. Log in using your MS Account credentials.

3. In the *Devices* area, tap *View all devices.*

4. Go through all your devices to make sure they all belong to you and that you have possession of them. *Remove* all others.

17.7 Apple ID and iCloud Security

In 2012 a well-known journalist had his Apple ID hacked, allowing the hacker full access to the victim's Apple ID, and through that, his iCloud account, including calendar, contacts, and email. This was accomplished not by traditional black hat hacking, but with a bit of social engineering. All the hacker needed was to discover the victim's birthdate and email address associated with his Apple ID. The hacker would look at an email to Apple that said something like, *I've forgotten my Apple ID password and would like to reset it. Here is my birthdate and my email address.* Using this information, the hacker could reset the Apple ID password. Once done, they could access the victim's iCloud website as if they were the actual victim.

I have had several clients whose Music.app/iTunes accounts have been compromised in a similar fashion. One was billed $1,400 in music purchases.

As of macOS 10.13, Two Factor Authentication (2FA) is mandatory for your Apple ID. Adding this security layer makes it extremely difficult for anyone to hijack your Apple ID and make fraudulent purchases.

Remember that every password and security measure can be broken. Your defense is to make it so difficult and time consuming to break that the hacker moves to an easier target. Also, most security questions can be accurately guessed or broken through social engineering: *What is your birthday? In what city did your parents marry? What is the name of your first pet?* etc. When security questions are based on something that you know, someone else can know that.

With Apple 2FA a code is sent to your previously verified devices whenever you sign into your Apple ID on the web to manage your account, purchase something from iTunes, App Store, or iBooks Store from a new (unknown) device or attempt to get Apple ID-related support from Apple. You are prompted to provide this 2FA code before the purchase or support can occur.

If your device has been stolen or lost, log into *https://appleid.apple.com* to remove that device from the verified device list, so that no 2FA code is sent to that device.

- Note: As of August 2016, NIST has stopped recommending Two-Factor Verification that involves SMS/text messaging as the second factor[13]. This is due to how easy it is to intercept the code. But *any* verification that is received via cellular signal (SMS, voice, etc.) is subject to the same vulnerabilities. Currently, the best solution is to use a digital token (a keychain-sized device that displays random number strings), or an *Authenticator* app, which serves the same purpose. At this time, Apple does not use digital tokens or an authenticator app but can provide verification codes to your phone.

---

[13] *https://pages.nist.gov/800-63-3/sp800-63-3.html*

17.7.1   Assignment: Enable Apple 2-Factor Authentication

Apple requires 2FA for all accounts. If you have upgraded from an older version of macOS or iOS that did not require 2FA, you enable 2FA in this assignment, then change your 2FA to have Apple directly send verification codes.

Setting Up Apple 2FA

1. Open Safari and go to *https://appleid.apple.com.*

2. Sign in with your Apple ID and password.

3. In the *Security* section, select the *Edit* button.

4. In the *Two-Factor Authentication* area, select enable *Two-Factor Authentication.*

5. Follow the brief on-screen instructions to complete Apple 2FA setup.

Using Apple 2FA

6. From now on, when accessing an Apple site, you may be prompted to provide Apple 2FA for access.

7. Open Safari to *https://appleid.apple.com.*

8. Enter your *Apple ID* and *password.*

9. On all your devices which are signed in with your Apple ID (including your mobile device), the *Apple ID Sign In Requested* dialog appears.

10. Tap *Allow*.

11. The code immediately appears onscreen.

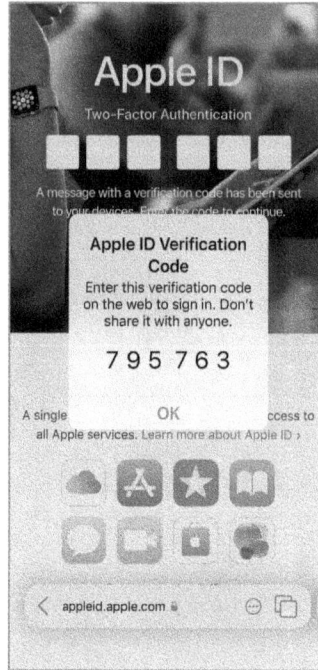

12. Enter this code into your Apple web page.

13. The site is now open for your use.

17.7.2 [Optional] Assignment: Remove A Device from Two-Factor Authentication

All devices (computers, iPads, iPhones, Apple TVs, etc.) on which you have signed into your Apple account receive *Apple ID Verification Codes* when an attempt is made to access your Apple account. Should one of your devices become lost, stolen, sold, or given to someone, the person who takes possession of it may be able to see your verification codes, presenting a security vulnerability.

To prevent this from happening, you must remove the device from your Apple ID device list.

In this assignment, you remove a device from your Apple ID device list.

● Note: Unless you really do wish to remove the device, skip this assignment.

5. Open a browser to *https://appleid.apple.com*.

6. At the prompt, enter your Apple ID email address and Apple ID password.

7. If you have set up 2-Factor Authentication, your devices display an alert that someone is attempting to access your account. Select *OK*. An *Apple ID Verification Code* appears.

8. Enter the *Apple ID Verification Code* in the alert window in your browser.

9. Scroll down to the *Devices* area.

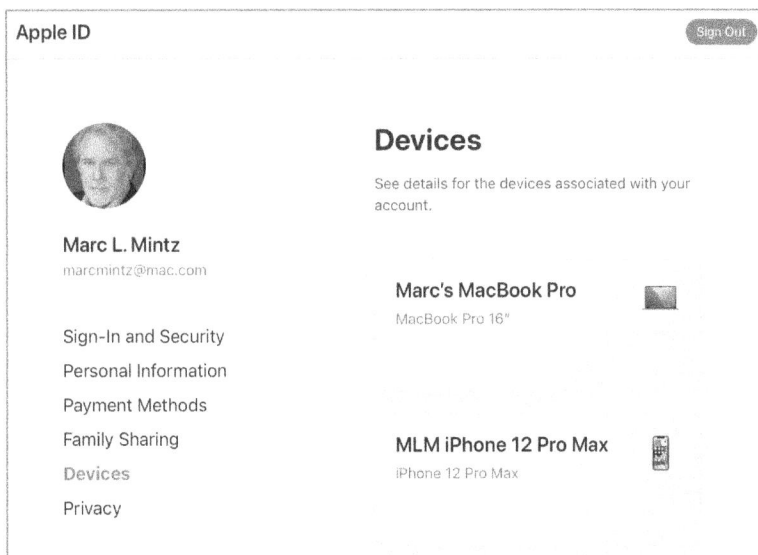

10. Tap on the device to be removed.

11. A pop-up window appears. Tap *Remove <device type>*.

12. The device is detached from your Apple ID and no longer receives Apple ID Verification Codes.

17.8   Apple Account Recovery and Legacy Contact

New with iOS 15 and macOS 12 is *Apple Account Recovery*. In the event you have lost the ability to access your iCloud data, Account Recovery helps you regain access when you can't reset your password. Some data cannot be recovered, such as keychain, Screen Time, and Health.

Also new with iOS 15 and macOS 12 and associated with Account Recovery is *Legacy Contact*. With Legacy Contact, you can specify someone of trust to have access to data in your account after your death.

17.8.1   Assignment: Configure Account Recovery

In this assignment, you set up account recovery, so that in the event you cannot access your Apple account and cannot change its password, you can still gain access.

- Prerequisite: All devices signed in with your Apple ID must be on the current OS (macOS 12 and higher, iOS 15 and higher, iPadOS 15 and higher).

1. Open *Settings* > your name > *Password & Security*. The *Password & Security* widow opens.

Trusted Phone Number

Used to verify your identity when signing in on a different device or browser.

2. With an iPhone, the *Trusted phone numbers* field will be populated with the device mobile phone number. If you would like to add an additional phone number, tap *Edit*, then add a phone number where you can receive either text or voice calls:

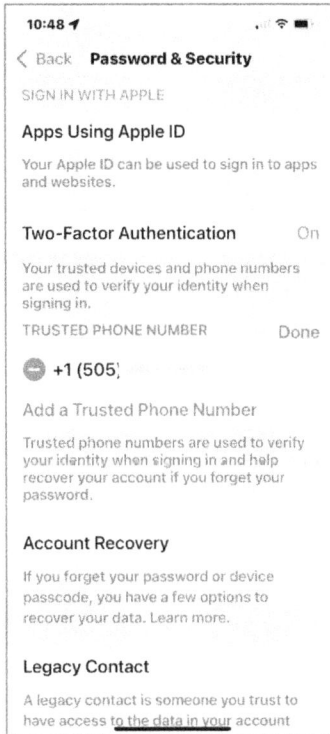

Account Recovery

Used by someone you trust to gain access to your data in the event you forget your account or device password.

- NOTE: By enabling *iCloud Data Recovery Service,* Apple can access your encrypted data, except for end-to-end encrypted data such as Keychain, Screen Time, and Health data

3. Tap *Settings > Your Name > Password & Security > Account Recovery >* tap *Add Recovery Contact.* The *Account Recovery Contact* window appears:

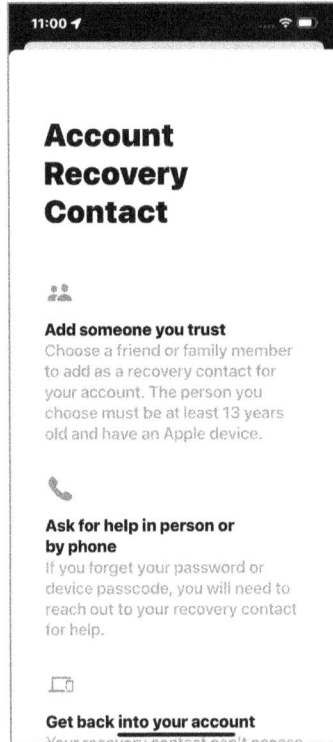

4. Scroll down, then tap *Add Recovery Contact* button.

5. Enter your Apple account password, tap *Next*, then follow the onscreen instructions.

Set Recovery Key

By setting a *Recovery Key*, you help secure your account from others by making it impossible to change your account password without entering the recovery key, or having access to another device already signed in with your Apple ID.

Enabling a recovery key also makes it possible to recover your data stored on Apple servers. Be aware that this also gives Apple access to all your data.

6. In the *Settings > Apple ID > Password & Security > Account Recovery* field, tap *Edit* button, then follow the onscreen instructions.

Digital Legacy Contact

After you have passed away, you can set up an account beneficiary who will have access to your Apple account and digital legacy.

7. In the *Settings > Apple ID > Password & Security >Legacy Contact* field, tap *Edit* button. The *Legacy Contact* window appears.

8. Tap *Add Legacy Contact*. The *Legacy Contact* window opens.

9. Tap *Add Legacy Contact*.

10. At the authentication window, tap *Use Password*, then enter your Apple ID password.

11. In the *Add Legacy Contact* window, tap *Choose Someone*, then follow the onscreen instructions to provide them account access upon your death.

12. *Exit* Settings.

17.9 Advanced Data Protection

Apple introduced Advanced Data Protection[14] as of iOS 16.2, iPadOS 16.2, and macOS 13.1. This is an option that provides for the highest level of security for iCloud data by enabling end-to-end data encryption between your device and Apples iCloud servers with the encryption keys stored only on your trusted devices and encrypting 23 data categories. Once enabled, not even Apple (or state actors, cyber-criminals, or even your government armed with search warrants) can read your data on iCloud.

Prior to the introduction of Advanced Data Protection, or with Advanced Data Protection disabled, the data you store on iCloud has *Standard Data Protection*. 14 data categories of iCloud data are end-to-end encrypted, but the keys are stored on Apple servers. This protects your data from standard hacking attacks but does not protect against targeted attacks or search warrants.

Data Category	Standard Data Protection		Advanced Data Protection	
	Encryption	**Key Storage**	**Encryption**	**Key Storage**
Apple Card Transactions	End-to-End	Trusted Devices	End-to-End	Trusted Devices

---

[14] *https://support.apple.com/en-us/HT202303*

Data Category	Standard Data Protection		Advanced Data Protection	
	Encryption	Key Storage	Encryption	Key Storage
Contacts	In Transit & On Server	Apple	In Transit & On Server	Apple
Calendars	In Transit & On Server	Apple	In Transit & On Server	Apple
Health Data	End-to-End	Trusted Devices	End-to-End	Trusted Devices
Home Data	End-to-End	Trusted Devices	End-to-End	Trusted Devices
iCloud Backup (including device and Messages backup)	In Transit & On Server	Apple	In Transit & On Server	Trusted Devices
iCloud Drive	In Transit & On Server	Apple	In Transit & On Server	Trusted Devices
iCloud Mail	End-to-End	Apple	In Transit & On Server	Apple
Maps	End-to-End	Trusted Devices	End-to-End	Trusted Devices
Memoji	End-to-End	Trusted Devices	End-to-End	Trusted Devices
Messaged in iCloud	End-to-End	Trusted Devices	End-to-End	Trusted Devices
Notes	In Transit & On Server	Apple	End-to-End	Trusted Devices
Passwords & Keychain	End-to-End	Trusted Devices	End-to-End	Trusted Devices
Payment Information	End-to-End	Trusted Devices	End-to-End	Trusted Devices
Photos	In Transit & On Server	Trusted Devices	End-to-End	Trusted Devices
QuickType Keyboard Learned Vocabulary	End-to-end	Trusted Devices	End-to-End	Trusted Devices
Reminders	In Transit & On Server	Apple	End-to-End	Trusted Devices

Data Category	Standard Data Protection		Advanced Data Protection	
	Encryption	Key Storage	Encryption	Key Storage
Safari	End-to-End	Trusted Devices	End-to-End	Trusted Devices
Safari Bookmarks	In Transit & On Server	Apple	End-to-End	Trusted Devices
Screen Time	End-to-End	Trusted Devices	End-to-End	Trusted Devices
Siri Information	End-to-End	Trusted Devices	End-to-End	Trusted Devices
Siri Shortcuts	In Transit & On Server	Apple	End-to-End	Trusted Devices
Voice Memos	In Transit & On Server	Apple	End-to-End	Trusted Devices
Wallet Passes	In Transit & On Server	Apple	End-to-End	Trusted Devices
Wi-Fi Passwords	End-to-End	Trusted Devices	End-to-End	Trusted Devices
W1 and H1 Bluetooth Keys	End-to-End	Trusted Devices	End-to-End	Trusted Devices

17.9.1 Assignment: Enable Advanced Data Protection

In this assignment, you enable Advanced Data Protection.

1. Verify that all of your trusted devices are updated to at least iOS 16.2, iPadOS 16.2, or macOS 13.1

9. Open *Settings > (User Name) Apple ID > iCloud > Advanced Data Protection.*

10. In the *Advanced Data Protection* windows, scroll down to select *Account Recovery > Set Up* button.

11. In the *Account Recovery* windows, if you do not yet have *Recovery Assistance* and *Recovery Key* sections completed, do so now, and then tap the *Done* button.

12. Back at the *Advanced Data Protection* window, tap the *Turn On* button.

13. In the *You will be responsible for your data recovery* pane, tap the *Review Recovery Methods* button.

14. At the prompt, enter your 28-character recovery key, then tap the *Continue* button.

15. At the prompt, enter the login password you use for this device.

16. At the *Advanced Data Protection is On* window, tap the *Done* button.

17. Tap the *Done* button.

18. In the *iCloud* window, scroll to the bottom to the *Access iCloud Data on the Web* field.

19. If you would like to be able to access your calendar, contacts, documents, mail, photos, notes, and reminders via a browser from your trusted devices, enable this field.

20. Exit *System Settings.*

17.10 Cybersecurity and Privacy Lessons Learned

☐ Facebook reported a 2020 profit of 10 million dollars, most of this using marketing information about you and other users.

☐ All social media sites should be hardened with 2FA and regular reviews of all security and privacy settings.

☐ You can use Google Takeout to automatically send you on a regular basis a comprehensive report of all it knows about you.

☐ By default, all devices registered with the same Apple ID receive 2FA Apple ID Verification Codes.

☐ New with macOS 12 and iOS 15 are *Apple Account Recovery* making it easier to regain control over your account once locked out, and *Legacy Contact,* where you can grant someone access to your account and digital assets following your death.

☐ New with macOS 13.1 and iOS 16.2 is *Advanced Data Protection*, expanding end-to-end encryption from14 to 23 data categories, and moving the encryption keys from Apples servers to your own trusted devices.

18 Apple Pay and Credit Cards

While money can't buy happiness, it certainly lets you choose your own form of misery.

–Groucho Marx[1]

What You Will Learn in This Chapter

- Escape from the epidemic of credit card theft
- How to configure your iPhone or iPad for Apple Pay
- How to use Apple Pay

What You Will Need in This Chapter

- [Optional] Apple Pay compatible debit, credit, or gift card.

18.1   The Epidemic of Credit Card Theft

In 2018, over $24 billion was lost due to card fraud worldwide. How can there be as many as 2 credit card breaches per adult in a year?

One answer is that merchants *love* to keep your credit card information in their greedy little hands. By storing your card information, it is effortless for the merchant to close a sale. With the credit card on record, the buyer has neither the time nor bother of having to search for the card while making a purchase.

This arrangement makes the merchant customer credit card database look like Fort Knox to a cyber thief. With enough resources and time, the thief can breach

---

[1] *https://en.wikipedia.org/wiki/Groucho_Marx*

the database, harvest millions of personal identity and credit card records, then turn Target, Home Depot, or any other merchant into their money machine!

There are strategies available to help avoid such problems. A key strategy is to primarily keep pieces of personal and card information on different servers. But even with such a strategy it is possible to breach all the data given adequate resources, especially if an insider is involved.

According to cyber security experts, the real answer lies in preventing the merchant from storing your data in a manner useable by anyone but the individual owner. Several organizations have come up with such a solution in the past few years (Apple Pay, Google Pay, and Samsung Pay). Apple is the first to do it right in the form of *Apple Pay*[2].

When using Apple Pay, the merchant never has access to your credit card number, expiration date, security code, or any other identifiable aspect of your card. All the merchant gets is a one-time-use code that confirms that you do indeed have a valid credit/debit/gift card, and that you (or at least the thumb used to press on the home button) are the rightful holder of said card. The merchant is authorized to charge to that card. But all the merchant has at the end of the transaction is data about the service/merchandise that was purchased, and a one-time-use code! There is nothing in the database worth stealing, and your card information remains hidden behind hardware encryption on your iOS device.

If you do not already have an Apple Pay compatible card, for the sake of security, it may be time to get one. A list of all Apple Pay compatible cards is available at: *https://www.creditcards.com/apple-pay.php*.

Now that you have your card, let's set up your device to use it with Apple Pay.

18.1.1   [Optional] Assignment: Configure Your Device for Apple Pay

In this assignment, you configure your Apple Pay compatible iOS device to use a credit card.

- Prerequisite: Possession of an Apple Pay compatible credit, debit, or pre-paid card.

---

[2] *https://en.wikipedia.org/wiki/Apple_Pay*

1. Pull out your bright and shiny debt-making sliver of plastic. You need to enter information regarding the card into your iOS device.

2. Tap *Settings* then scroll down to tap *Wallet & Apple Pay.*

3. Tap Add Card.

4. If this is your first time visiting this setting, the *Apple Pay* screen appears. Tap *Continue.*

5. The *Card Type* screen appears. Tap *Debit or Credit Card.*

6. Tap *Continue.*

7. At the *Add Card* screen, the rear camera activates.

8. Position a credit, debit, or gift card within the top frame. The card details (holder name and card number) are automatically captured and displayed. Select *Next.*

9. In the *Card Details* screen, verify the information, then tap *Next.*

10. In the second *Card Details* screen, enter the *Expiration Date* and *Security Code,* then tap *Next.*

11. In the Terms and Conditions screen, select Agree.

12. The verification process may take a minute or two. When complete, you are returned to the *Wallet & Apple Pay* screen, with your new Apple Pay card listed.

18.1.2   [Optional] Assignment: Use Apple Pay In Stores

In this assignment, you make a purchase using Apple Pay.

- Prerequisite: Completion of the previous assignment.

1. Go to a brick-and-mortar store that accepts Apple Pay. Close your eyes, take a few steps, you are bound to trip over one. Or if you want to play it safe, look for the Apple Pay logos:

2. When you are at the checkout register and ready to make a payment, hold your iOS device within an inch of the contactless terminal and:

   - If your iOS device is locked, double tap the *Home Button* to allow access.

   - If your iOS device is unlocked, have your finger touching the *Touch ID* home button (do not press down or tap).

3. Your default card appears on your screen.

   - If you wish to pay with the default card, just hold there a second, then skip to the next step.

   - If you prefer to pay with a secondary card, tap the image of the default card. A list of all your Apple Pay cards appear, tap one.

4. Some terminals prompt you to select *Credit* or *Debit.* Apple's recommendation is to always select *Credit,* as older terminals may not work when selecting *Debit.* I know, go figure.

Ahhhh… Didn't that feel wonderful? Retail therapy *and* no chance of identity or credit card theft.

18.2 Cybersecurity and Privacy Lessons Learned

☐ Measuring by dollars lost, most credit card theft is due to vulnerabilities at the merchant and credit card processing groups, due to their desire to maintain full records of credit purchases, including all card information.

☐ Currently the best solution to credit card theft and fraud is to prevent the merchant from having access to credit card data. This has been implemented with Apple Pay, Google Pay, and Samsung Pay.

19 Medical and SOS Emergency

'Emergencies' have always been the pretext on which the safeguards of individual liberty have been eroded.

–Freidrich August von Hayek[1], co-recipient of the 1974 Nobel Memorial Prize in Economic Sciences

What You Will Learn in This Chapter

- Provide your medical and emergency contact information in an emergency
- Send an SOS emergency call

What You Will Need in This Chapter

- No additional resources required.

19.1 Medical Emergency

In the event of an emergency, your iOS device can provide critical health and emergency contact information to first responders, even when locked.

19.1.1 Assignment: Configure Health

In this assignment you configure your emergency health information.

1. On your iOS device, open *Settings > Health > Health Details*. The *Health Details* screen opens.

---

[1] *https://en.wikipedia.org/wiki/Friedrich_Hayek*

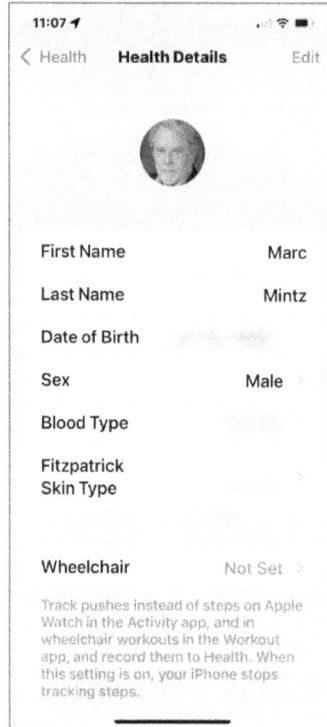

2. Tap *Edit,* then complete all the requested fields. When done, tap *< Health* to go back.

3. Tap *Medical ID.* The *Medical ID* screen opens.

4. Tap *Edit,* then complete all the requested fields.

5. Scroll down to the *Emergency Access* area, then enable *Show When Locked.* This allows first responders to view your Medical ID information on your locked device by tapping *Emergency > Medical ID.*

6. You may want to enable *Share During Emergency Call.* This sends your encrypted Medical ID information and GPS location to Apple. If the *Enhanced Emergency Data Service* is available in your area, Apple forwards your information to a partner for delivery to emergency services.

7. When complete, tap *Done,* then tap *< Health.*

19.1.2   Assignment: Test Health

In this assignment you test that your Medical ID information is available to first responders.

- Prerequisite: Completion of the previous assignment.

1. Lock your device screen.

2. Wake your device. Do not use Touch ID or Face ID to logon. Instead, acting as a first responder, swipe up to access the numeric keypad screen.

3. At the bottom left of the screen, tap *Emergency*. The *Emergency Call* screen appears.

4. At the bottom left corner tap *Medical ID*. Your Medical ID information appears.

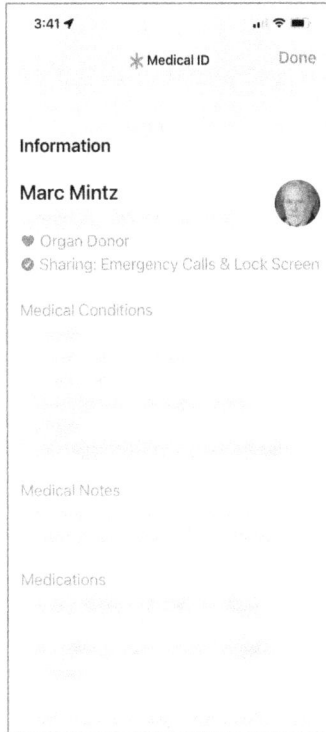

5. To exit, tap *Done,* then *Cancel.*

19.2 Emergency SOS

iOS has integrated *Emergency SOS*, which automatically calls the local emergency phone number. This call is made somewhat covertly as you do not need to dial a number, but instead press a combination of side buttons.

You also have the option to add emergency contacts. After the Emergency SOS call ends, your iPhone alerts your designated emergency contacts with a text message, including your current GPS location, and for a period any changes in your location.

19.2.1 Assignment: Configure Emergency SOS

In this assignment, you configure *Emergency SOS*.

- Prerequisite: Completion of the previous assignment.

1. On your iPhone, open *Settings > Emergency SOS.* The *Emergency SOS* main screen appears.

2. By default, by pressing and holding the side button and either volume button, the emergency call is made. *Enable Call with Side Button* to allow pressing the side button 5 times to also make an emergency call.

3. *Enable Auto Call.* This removes a verification step and speed making the emergency call. You still have a 3-second countdown timer in which you can cancel the call.

4. To exit, tap *<Settings.*

19.2.2   Assignment: Test Emergency SOS

In this assignment, you test your Emergency SOS system.

1. From any screen on your device, quickly tap the side button 5 times.

2. If working, you see a countdown timer starting at 3. Tap *Stop > Stop Calling* before the countdown reaches 0 to prevent your device from calling emergency services.

3. From any screen on your device, press and hold the side button and either volume button for 2 seconds.

4. Your device should immediately start calling emergency services. Tap *Stop* to stop the call.

5. Your device should immediately start contacting your emergency contacts. Tap *Stop* to stop the process.

6. Your *Medical ID* information appears. Tap *Done* to exit.

19.3 Medical and SOS Emergency Lessons Learned

☐ You may store your emergency health information on your iOS device in *Settings > Health > Health Details.*

☐ Your iOS device can notify authorities and friends of an emergency SOS. This is configured in *Settings > Emergency SOS.*

20 When it is Time to Say Goodbye

Don't cry because it's over. Smile because it happened.

–Dr. Seuss[1]

What You Will Learn in This Chapter

- Prepare an iOS device for sale or disposal

What You Will Need in This Chapter

- No additional resources required.

20.1   Prepare A Device for Sale

The time comes when all good things must come to an end. This is just as true for your beloved iPhone or iPad. But your device holds all your documents, passwords, pictures, web browsing history, and other items you do not want someone else to see. Even if you are tossing your damaged device in the trash, there is the real probability that someone can find it and harvest all your data.

Your mobile device is connected with the Apple environment on several layers, all of which must be disconnected. These include:

- Music.app/iTunes
- iCloud
- Messages

---

[1] *https://en.wikipedia.org/wiki/Dr._Seuss*

Once your device is disconnected from the Apple environment, the drive itself must be securely erased or destroyed. If the device will be reused, iOS or iPadOS should be reinstalled.

20.1.1 [Optional] Assignment: Transfer Information to Your New Device

If your older device is still fully functioning, after turning on your new device you can transfer all information from the old to new device by following the onscreen *Quick Start* instructions.

If your older device is not fully functioning, or you do not have access to it, you can use iCloud, iTunes, or Finder to transfer information from the last backup of the older device to the new device.

20.1.2 [macOS] Assignment: Remove Firmware Password

20.2 [Android, macOS and Windows] Secure Erase Storage Devices

20.2.1 [Android, Chrome, macOS, Windows] Assignment: Secure Erase Boot Drive

20.2.2 [Android, Chrome, macOS, Windows] Assignment: Secure Erase a Non-Boot Drive

20.2.3 [Windows] Assignment: Create a Bootable Parted Magic USB Drive

20.2.4 [Windows] Assignment: Secure Erase An Encrypted SSD With Recovery Drive

20.2.5 [Windows] Assignment: Secure Erase an Unencrypted SSD With Parted Magic

20.3 [Android, Chrome OS, macOS, and Windows] Install OS

20.3.1 [Android, Chrome OS, macOS, and Windows] Assignment: Install macOS 13

20.3.2 [Optional] Assignment: Prepare a Device for Sale or Disposal

Early iOS devices (original iPhone, iPhone 3G, iPod touch, and iPod touch 2nd generation) erase by overwriting data in memory. Erasing these devices may take an hour or more. All iOS devices that support iOS 9 and higher use hardware encryption. Erasing these devices is a matter of removing the encryption key, taking only minutes.

In this assignment, you prepare a device for sale or disposal, including a secure erase on your device.

- **Warning**: Unless you really do wish erase your device, skip this assignment.

- **Warning**: Completing this assignment will erase your device. Erasing your device is a non-reversible process. Before continuing, perform both an iCloud and iTunes/Finder backup of the device.

Back up your iOS device

1. Back up to iCloud.

2. Back up to computer.

Unpair your Apple Watch

3. If you have paired your Apple Watch to your iPhone, unpair it[2].

Sign out of iCloud and the iTunes & App Store

---

[2] *https://support.apple.com/kb/HT204568*

4. Go to *Settings > Your Name*.

5. Scroll to the bottom of the *Apple ID* screen, then select *Sign Out.*

6. At the prompt, enter your *Apple ID password,* then tap *Turn Off.*

Verify removal from your Apple account

Verify that the device has been removed from your Apple account from the step above.

7. Open a browser to *https://appleid.apple.com/.* Log in with your Apple ID and password.

8. Scroll down to the *Devices* area.

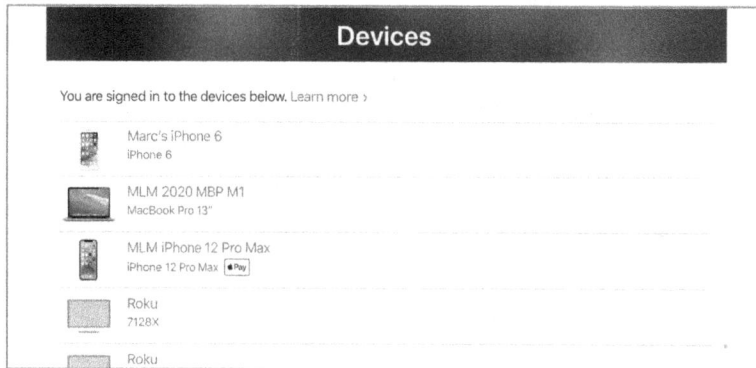

9. If the device is not found, all is good, and you can skip to the next step. If the device is displayed:

    a. Select the device to be removed from your account.

    b. From the pop-up menu or screen, select *Remove from account.*

10. Exit from *Settings*.

Unlock the iPhone

● Note: Skip if the device is an iPad.

● Note: A device can only be unlocked if it is paid in full.

If the device is an iPhone, unlocking it allows the next owner to use it on another carrier.

11. Contact your current carrier and request the iPhone be unlocked. Some carriers have a webpage to unlock your device.

Deregister iMessage

12. If you are switching to a non-Apple device, deregister iMessage[3].

Erase all Content and Settings

13. Select *Settings > General > Reset*, then tap *Erase All Content and Settings*.

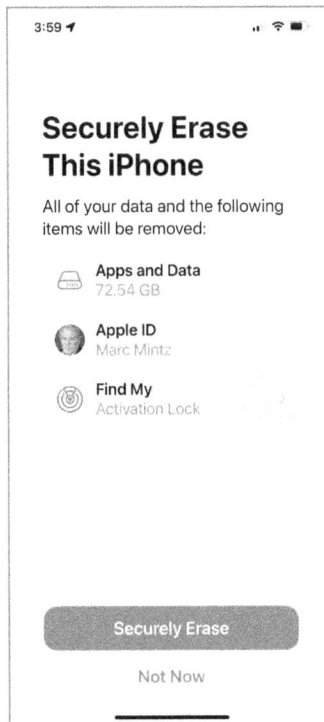

14. In the *Securely Erase This iPhone* screen, tap *Securely Erase*.

15. Once complete, your device may restart, taking you to the login screen.

Remove the SIM card

---

[3] *https://support.apple.com/kb/HT204568*

- Note: If your device is an iPad or an iPhone with built-in (non-removable) SIM card, skip this step.

16. If your iPhone uses a SIM card, remove, then destroy it.

Done!

20.4   When it is Time to Say Goodbye Lessons Learned

☐ Before selling, giving away, or even trashing a device, it must be properly processed.

☐ The factory reset an iOS device, go to Settings > General > Reset > Erase All Content and Settings.

21 Miscellaneous

The nice thing about standards is that you have so many to choose from.

–Andrew S. Tanenbaum[1]

What You Will Learn in This Chapter

- Configure date and time settings
- Comply with National Institute of Standards and Technology (NIST)
- Comply with United States Computer Emergency Readiness Team (US-CERT)
- Comply with International Organization for Standardization (ISO)

What You Will Need in This Chapter

- No additional resources required.

21.1 Date and Time Settings

There are several reasons it is critical to keep your computer date and time accurate:

- If you are on a network with other computers, or use a server, if your clock is off by more than a few minutes, you may be blocked as the other systems see this as a potential *man in the middle* attack.

---

[1] *https://en.wikipedia.org/wiki/Andrew_S._Tanenbaum*

- Should your device become compromised by malware or a criminal, it is important to know the exact moment the penetration occurred.

- You do not want to be late to your Aunt Rose's dinner party. Noodle Koogle is best right out of the oven.

Fortunately, most devices with internet connectivity include the ability to automatically synchronize with atomic clocks around the world using the Network Time Protocol[2] (NTP).

21.1.1   Assignment: Configure Date & Time

By default, your iPhone and iPad are configured to automatically synchronize with the NTP server of your wireless carrier. But if your device is not using a wireless carrier, you need to manually set date and time. Also, should you be traveling into a different time zone, it can take 24 hours for your device to acknowledge the new zone.

In this assignment, you manually set date and time.

1. Open *Settings.app > General > Date & Time.* The *Date & Time* screen appears.

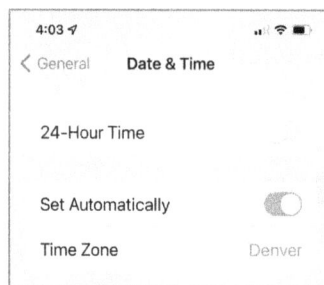

4:03 ✓		.ıl 🌢 ∎
‹ General	**Date & Time**	
24-Hour Time		
Set Automatically		⬤
Time Zone		Denver

2. If you need to manually set date and time, turn off *Set Automatically.*

3. Configure to your taste, then exit *Settings.*

---

[2] *https://en.wikipedia.org/wiki/Network_Time_Protocol*

21.2 National Institute of Standards and Technology (NIST)

NIST, part of the U.S. Department of Commerce, was established by Congress in 1901 to improve our measurement systems because substandard measurement systems were impeding U.S. economic growth. This NIST mission continues into the 21st-century by developing standards of measurement at the nanoscale through galactic scale.

NIST also is involved in developing standards for computer and IT systems. Some organizations–particularly healthcare, financial, and legal–base much of their cybersecurity measures on the *NIST SP 800-171 Protecting Controlled Unclassified Information in Nonfederal Systems and Organizations*[3].

As of the time of this writing, NIST had just released *Draft NIST Special Publication 800-179 Guide to Securing Apple macOS 10.12 Systems for IT Professionals: A NIST Security Configuration Checklist*[4]. This publication details how NIST recommends configuring the security of macOS 10.12.

NIST also publishes the *NIST SP 800-53 Security and Privacy Controls for Information Systems and Organizations*[5]. The current release is revision 5, released September 2020. This document was originally intended for standards and guidance for federal organizations. However, with the most current release, the name "Federal" was removed from the title in recognition that the guidelines are appropriate for all organizations.

Following the steps outlined in *Practical Paranoia*, you can likely pass a NIST-based security audit. That said, if your organization is held to NIST protocols, you are held accountable for knowing them from the NIST perspective, and reading their guide is mandatory.

---

[3] *https://nvlpubs.nist.gov/nistpubs/SpecialPublications/NIST.SP.800-171r1.pdf*

[4] *https://csrc.nist.gov/CSRC/media/Publications/sp/800-179/rev-1/draft/documents/sp800-179r1-draft.pdf*

[5] *https://csrc.nist.gov/publications/detail/sp/800-53/rev-5/final*

21.2.1 [Optional] Assignment: NIST Department of Defense Configuration

NIST maintains Department of Defense IT checklists[6] for operating systems, servers, and applications. If your organization must comply with DoD protocols, this is a perfect resource. There are currently 23 pages of configuration settings for iOS[7] and iPadOS–with most of the settings only available when the device is configured in a *Mobile Device Management (MDM)* environment. Although MDM configuration can be done without MDM software by using Apple Configurator, either option is beyond the scope of the book and course. That said, once you have learned either your MDM software or Apple Configurator, all your iOS and iPadOS devices can be secured remotely and without involving your users.

21.2.2 [Optional] Assignment: NIST-Specific Electronic Health Records on Mobile Devices

In July 2015, NIST released their *NIST Cybersecurity Practice Guide, Special Publication 1800-1: Securing Electronic Health Records on Mobile Devices*[8, 9, 10,]

---

[6] *https://nvd.nist.gov/ncp/repository?sortBy=modifiedDate%7Cdesc*

[7] *https://nvd.nist.gov/ncp/checklist/967/download/6433*

8 *Executive Summary:*
   https://nccoe.nist.gov/sites/default/files/library/sp1800/hit-ehr-nist-sp1800-1a-draft.pdf

[9] *Approach, Architecture, and Security Characteristics:*
   https://nccoe.nist.gov/sites/default/files/library/sp1800/hit-ehr-nist-sp1800-1b-draft.pdf

[10] *How-To Guide:*
   https://nccoe.nist.gov/sites/default/files/library/sp1800/hit-ehr-nist-sp1800-1c-draft.pdf

[11, 12]. This document is mandatory reading if your work includes supporting anyone within the healthcare community.

21.3 United States Computer Emergency Readiness Team (US-CERT)

US-CERT[13], part of the Department of Homeland Security, began in early 2000 as a response to Federal Government networks experiencing an alarming number of cyber breaches. Congress created the Federal Computer Incident Response Center (FedCIRC) at the General Services Administration as a centralized hub of coordination and information sharing between federal organizations. With the creation of the Department of Homeland Security in 2002, Congress transferred US-CERT responsibilities to FedCIRC. In 2003, FedCIRC was renamed US-CERT, and its mission was expanded to include providing boundary protection for the federal civilian executive domain and cybersecurity leadership.

This shared responsibility has evolved over time to make US-CERT a trusted partner and authoritative source in cyberspace for the Federal Government, State, Local, Tribal and Territorial governments, private industry, and international organizations.

21.4 International Organization for Standardization (ISO)

---

[11] *Standards and Controls Mapping: https://nccoe.nist.gov/sites/default/files/library/sp1800/hit-ehr-nist-sp1800-1d-draft.pdf*

[12] *Risk Assessment and Outcomes: https://nccoe.nist.gov/sites/default/files/library/sp1800/hit-ehr-nist-sp1800-1e-draft.pdf*

[13] *https://www.us-cert.gov/*

ISO[14] is an independent, international, standards organization. Many industries–in particular those dealing with banking, SEC-compliance, financial advisors, and SEC-compliance–rely on one of the many ISO standards. The ISO 27001[15] is the primary standard covering IT security.

Although many books are available to assist with meeting ISO 27001 compliance, it is best to first purchase the ISO 27001 book[16], which details the compliance matrix. This is only available from the ISO.

21.5   Vehicle Infotainment Systems

In August 2016, the FTC[17] and US-CERT[18] released an alert regarding securing personal information when using rental vehicles.

When using a rental vehicle with smart features, it is likely these tools are storing data about your travel, including full GPS mapping of your path, your speed at each point along the path, where you stopped, and for how long. Unfortunately, it is unlikely you can do much about this as most of the systems do not allow erasing this data.

When connecting a digital device such as a smartphone or tablet to a vehicle, it is possible the vehicle infotainment system automatically imports personal data such as contacts, your phone number, text messages and emails.

This information is automatically saved in the infotainment memory, and may be viewed by the rental agency personnel, or the next renter. Some of this information may be used for identity theft, or other criminal activity.

To secure your personal data in a rental vehicle:

---

[14] *https://www.iso.org*

[15] *https://www.iso.org/isoiec-27001-information-security.html*

[16] *https://www.iso.org/standard/54534.html*

[17] *https://www.consumer.ftc.gov/blog/what-your-phone-telling-your-rental-car*

[18] *https://www.us-cert.gov/ncas/current-activity/2016/08/30/FTC-Releases-Alert-Securing-Personal-Information-When-Using-Rental*

- **Do not connect your digital device to the infotainment system.** Instead of connecting to the infotainment USB port, connect through a cigarette lighter USB adapter for charging. Some USB connections automatically transfer data upon connection.

- **Configure permissions.** If you do want or need to connect your mobile device to the infotainment system, the system may present a permissions screen to specify or limit what data is transferred. If so, transfer only your music list, not contacts, messages, emails, etc.

- **Delete all data upon returning the rental vehicle.** Prior to returning the rental vehicle, check the infotainment system settings for the list of devices it is paired with. Locate your device and delete it from the list.

21.6 Emergency Location Service

As of 2018, Android and iOS include *Emergency Location Service (ELS)* for Android 4.0 and higher and iOS 12 and higher. Apple calls this service *Instant 911 Location Sharing.* When an Android or iOS user calls 911, their device provides location information directly to the emergency service without going through Google or Apple servers. Using a combination of GPS, Wi-Fi, mobile networks, and sensors, ELS allows emergency service center to locate the caller within 121 feet.

21.7 Strong Hacking Protection

New to iOS 12 is additional hacking protection. Several companies have developed brute force hacking software for iOS, supposedly restricted to law enforcement. However, these same techniques can be used by criminals.

To use brute force attack, it is necessary to access the iOS system using a Lightning cable to connect with the attacker's computer.

With iOS 12 and higher, if the device has not been unlocked for an hour, the Lightning port is restricted to charging only–it turns off the Lightning port data channels. Prior to iOS 12 the same restriction took place, but only after 7 days of locked state.

21.8 Cybersecurity and Privacy Lessons Learned

☐ Maintaining accurate time on your device is critical for security.

☐ iOS devices have the built-in ability to synchronize with the atomic clocks participating in the Network Time Protocol (NTP), as well as local network servers.

☐ National Institute of Standards and Technology (NIST), part of the US Department of Commerce, is a major researcher and developer of cybersecurity and privacy best practices.

☐ United States Computer Emergency Readiness Team (US-CERT), part of the Department of Homeland Security, is tasked with developing standards of cybersecurity for federal government agencies, as well as non-governmental organizations that work with federal agencies.

☐ International Organization for Standardization (ISO), an independent international organization, creates standards for many industries including financial.

☐ When using a rented vehicle, do not connect your digital devices to the infotainment system. Instead, connect to a cigarette lighter USB adapter for charging.

22 The Final Word

If you followed each of the steps outlined in this book, your computer is secured to a level the NSA requires for most of its own staff. Although this will not prevent all the bad guys from stealing your computer, it prevents them from accessing your data. And since you have at least one current backup at the home or office, and one on the Internet, you still are in possession of the items with *real* value: your data and peace of mind.

22.1 Additional Reading

"Security Recommendations to Prevent Cyber Intruders." US-CERT Alert (TA11-200A). July 2011. US-CERT. <https://www.us-cert.gov/ncas/alerts/TA11-200A

iOS 16 Security Checklist

Included below is the checklist used by Mintz InfoTech, Inc. consultants when performing security checks for our clients' systems. You should use this same checklist to ensure your own system is fully hardened.

OS Version

- ☐ Verify iOS is up to date (1.14.1)

Data Loss

- ☐ Automatic backup to Finder/iTunes computer in use (2.1.2)
- ☐ Integrity test backup to computer (2.1.3)
- ☐ Automatic iCloud backup in use (2.1.8)
- ☐ Integrity test iCloud backups (2.1.9)

Passwords

- ☐ Verify all passwords for websites and services are strong (3.2.1)
- ☐ Verify simple (6 number) passcode created, with *Erase Data on Failed Passcode Attempts* enabled (3.2.2-3.2.3)

Or

- ☐ Complex (15 or more character) passcode created (3.2.2)
- ☐ Verify Require Passcode set to a short period (3.3.4)
- ☐ All challenge questions and answers securely recorded (3.4, 3.4.1)
- ☐ 2-Factor Authentication in use for all available sites and services, using either a 2FA app or 2FA USB key (3.8-3.8.2)
- ☐ Synchronize passwords across devices and browsers with Bitwarden or iCloud Keychain (Note: Not for use in secure facilities) (3.5-3.5.1, 3.8-3.8.2)

Optional

☐ Activate iCloud Keychain Synchronization (Note: Not for use in secure facilities) (3.5.1)

☐ Bitwarden installed and in use to securely create, store, retrieve passwords and 2-Factor Authentication (3.6-3.6.2)

☐ Verify Password Policies are in place (3.7-3.7.2)

System and App Updates and Privacy

☐ Verify all OS updates installed (4.1.1)

☐ Verify all application updates installed (4.1.1, 4.1.2)

☐ Verify Apps and Updates configured for Automatic Downloads (4.1.1)

☐ Verify *Settings > Privacy > Location Services* is configured to your taste, with acceptable privacy settings (4.2.1)

☐ Verify *Settings > Privacy > Tracking* is configured to your taste, with acceptable privacy settings (4.2.1)

☐ Verify *Settings > Privacy* settings for every function are configured to your taste, with acceptable privacy settings (4.2.1)

☐ Verify *Settings > Privacy > Apple Advertising > Personalized Ads* disabled (4.2.1)

☐ Verify *Settings > Privacy > Record App Activity > Record App Activity* enabled (4.2.1)

User Accounts

Optional

☐ Verify Screen Time configuration (5.3.2)

Device Hardware

☐ Verify SIM Card password assigned and securely recorded (6.7, 6.7.1)

☐ Know about camera and microphone recording indicator (6.8.1)

Sleep and Screen Saver

☐ Verify configuration of Auto-Lock (7.1.1)

☐ Verify configuration of Lock Screen Notifications (7.2.1)

☐ Verify configuration of Do Not Disturb mode (7.3.1)

☐ Configure Analytics & Improvements to your taste

Malware

☐ There is currently no need for anti-malware on iOS

Optional

☐ If using Outlook.com for email, verify Safelinks is active (8.4.1)

☐ Enable Lockdown Mode (8.6)

Firewall

☐ There is currently no need for a firewall on iOS (9.1)

Lost or Stolen Device

☐ Verify Find My iPhone/iPad active (10.1.1)

☐ Verify Enable Send Last Location (10.1.1)

Local Network

☐ Use WPA3 with AES for all Wi-Fi networks (11.3, 11.4.1, 11.4.2)

☐ Verify strong password in use for Wi-Fi

☐ Securely record all network passwords

☐ No Ethernet hubs in use, only Ethernet switches (11.4)

☐ Power-cycle modems and routers monthly (11.6.1)

☐ Verify firmware for modems and routers is up to date (11.6.1)

☐ Verify DNS settings on modems and routers (11.6.1)

☐ Verify Port Forwarding settings on modems and routers (11.6.1)

☐ Verify DMZ settings on modems and routers (11.6.1)

☐ Verify only known Bluetooth devices are connected (11.7.1)

☐ Pair Bluetooth devices in a secure area, away from possible eavesdroppers (11.7.2)

☐ Do not authorize Bluetooth connections unless you are trying to pair it for the first time (11.7.2)

☐ Remove lost and unrecognized Bluetooth devices from the paired device list (11.7.3)

Optional

☐ Enable MAC Address filtering to Limit Wi-Fi Access (11.5-11.5.1)

☐ Install RADIUS authentication for Wi-Fi and Ethernet access (11.1-11.2)

Web Browsing

☐ Never enter sensitive information in an HTTP page, only on an HTTPS page (12.1)

☐ Use only high privacy browser (12.2-12.2.1)

☐ Review browser security settings (12.2.1, 12.2.2)

☐ Configure DuckDuckGo as the default search engine (12.4, 12.4.1, 12.4.2)

☐ Do Not Track enabled (12.2.1, 12.2.2)

☐ Block 3rd-party Cookies enabled (12.2.1. 12.2.2)

☐ Fraudulent Website Warning enabled (12.2.1, 12.2.2)

☐ User educated on web scams (12.7, 12.9)

☐ Check for account breaches at haveibeenpwned.com (12.12, 12.12.1, 12.12.2)

☐ Block Fingerprinting enabled (12.8, 12.8.1)

☐ Turn off Apple Personalized Ads (12.13-12.13.2)

Optional

☐ Install the Onion browser (12.10)

Email

☐ User educated on how to recognize phishing attacks (13.2)

☐ Use either TLS or HTTPS for all email (13.3-13.5.1)

☐ Mail Privacy Protection enabled (13.6-13.6.1)

☐ Create SPF, DKIM, and DMARC records (13.11-13.11.4)

☐ Enable email 2-Factor Authentication (13.12)

Optional

☐ Email aliases and Hide My Email in use 13.7-13.7.3)

☐ ProtonMail account in use (13.8-13.8.3)

☐ Paubox account in use (13.4.5)

Documents

☐ Know how to encrypt within Apple Pages, Numbers, and Keynote (14.1, 14.2, 14.2.1, 14.2.2)

☐ Know how to encrypt a PDF document (14.3-14.3.2, 14.3.3)

☐ Know how to encrypt a folder for cross platform use with zip (14.5-14.5.2, 14.5.3)

☐ Know how to remove sensitive Exif data from images and audio files (14.6-14.6.2)

☐ Know how to remove sensitive metadata from Microsoft Office files (14.6.3)

☐ Know how to view and remove document and object metadata in pdf (14.6.6-14.6.8)

Voice, Video, And Instant Message Communications

☐ Signal used for secure voice, video, and text communications (15.1-15.3.4)

Internet Activity

☐ VPN in use at all times (16.1-16.3.2)

☐ Secure DNS in use at all times (16.6-16.6.3)

☐ Verify DNS leak test (16.6.1-16.6.3)

☐ Verify Private Relay in use (16.7.1)

Social Media, Apple ID and iCloud

☐ Verify only strong, unique passwords used for social media sites

☐ Verify all social media passwords are securely recorded

☐ Verify 2-Factor Authentication for all social media sites

☐ Review all Facebook security and login settings (17.3-17.3.2)

☐ Review Off-Facebook data (17.3.3)

☐ Review LinkedIn account security (17.4.1)

☐ Download a *Get a copy of your data* from LinkedIn (17.4.2)

☐ Review Google sign-in and security settings (17.5.1)

☐ Download your Google Takeout (17.5.2)

☐ Review Microsoft account security (17.6, 17.6.1)

☐ Review all devices linked to your Apple ID (17.7, 17.7.2)

☐ Review your Apple Account Recovery settings (17.8, 17.8.1)

☐ Enable Advanced Data Protection (17.9, 17.9.1)

Apple Pay and Credit Cards

Optional

☐ Review Apple Pay settings and attached credit cards (18.1.1)

Medical and SOS Emergency

☐ Review your Health Details settings (19.1.1, 19.1.2)

☐ Review your SOS Emergency configuration (19.2.1-19.2.2)

When It Is Time to Say Goodbye

☐ Know how to backup to iCloud (20.1, 20.1.1)

☐ Know how to backup to computer (20.1-20.1.2)

☐ Know how to unpair from your Apple Watch (20.3.2)

☐ Know how to sign out of iCloud and App Store (20.3.2)

☐ Know how to verify removal from your Apple account (20.3.2)

☐ Know how to deregister iMessage (20.3.2)

☐ Know how to erase the device (20.3.2)

☐ Know how to remove and destroy the SIM card (20.3.2)

Miscellaneous

- ☐ Verify date and time set correctly (21.1.1)

- ☐ Know not to connect mobile device to a rental vehicle infotainment system (21.5)

Revision Log

6.0 20230111

- Full synchronization with Practical Paranoia macOS 13.

6.0 20221230

- Minor edits

6.0 20221228

- Minor edits

6.0 20221214

- Added 17.9 Advanced Data Protection

6.0 20221125

- Minor edits

6.0 20221030

- Updated for recent changes in iOS 16
- Synchronized chapter, section, and assignments with other Practical Paranoia v6 books

Acronyms

2FA–*Two-Factor Authentication.* The use of a secondary source (SMS message, email, authenticator app, authenticator key) for authentication in addition of username and password.

ACSP–*Apple Certified Support Professional.* An Apple certification required as part of acceptance into the Apple Consultants Network.

ACTC–*Apple Certified Technical Coordinator.* Currently, the highest-level Apple certification.

AES–*Advanced Encryption Standard.* An algorithm for the encryption of electronic data.

BAA–*Business Associate Agreement.* A US legal document required of some persons and organizations who perform services for HIPAA-covered entities (health organizations).

CCC–*Carbon Copy Cloner.* A macOS-specific backup software.

CEO–*Chief Executive Officer.* The person who thinks they are the highest-ranking in an organization.

CIO–*Chief Information Officer.* The person who actually controls an organization, even if they typically report to the CEO.

CISPA–*Cyber Intelligence Sharing and Protection Act.* Allows sharing of internet traffic between the US Government and technology and manufacturing organizations.

DKIM–*DomainKeys Identified Mail.* An email authentication method designed to detect forged sender addresses.

DMARC–*Domain-based Message Authentication, Reporting, and Conformance.* An email authentication protocol, designed to give domain owners the ability to protect their domain from unauthorized use.

DMZ–*Demilitarized Zone.* A physical or logical subnetwork that contains all the external-facing services to an untrusted network, typically the internet.

DNS–*Domain Name Server* or *Domain Name System*. A hierarchical and decentralized naming system for computers, services, and other resources connected to a local network or internet.

DSL–*Digital Subscriber Line*. A technology for high-speed digital data over standard phone lines.

Exif–*Exchangeable image file format*. A standard for storing metadata for image and sound files.

FTC–*Federal Trade Commission* (US). A US antitrust and consumer protection agency.

GB–*Gigabyte*. A unit of information equal to 2^{30} bytes or approximately one billion bytes.

GPG–*GNU Privacy Guard*. Free cryptographic software.

HDD– *Hard Disk Drive*. A type of storage device containing rigid rotating platters.

HIPAA–*Health Insurance Portability and Accountability Act*. A US law that restricts access to individuals' private medical information.

HTTP–*Hypertext Transport (or Transfer) Protocol*. The data transfer protocol used on the World Wide Web.

HTTPS–*Hypertext Transfer Protocol Secure*. The encrypted data transfer protocol used on the World Wide Web.

iOS–The name of the operating system used by Apple iPhones.

IoT–*Internet of Things*. The network of objects connected to the internet, not normally viewed as computers or telecommunications. Examples are nanny cams, smart doorbells, etc.

IP–*Internet Protocol*. A standardized protocol for the transmission of digital data over both local networks and the internet.

iPadOS–The name of the operating system used by Apple iPads.

ISP–*Internet Service Provider*. A company that provides subscribers with access to the internet.

ISO–*International Organization for Standardization*. An international standard-setting body composed of representatives from various national standards organizations.

IT–*Information Technology*. The study or use of computers or telecommunications for storing, retrieving, and sending information.

LED–*Light Emitting Diode*. A semiconductor which glows when voltage is applied.

MAC–*Media Access Control address*. A unique identifier assigned to a network interface controller, typically used with computers, printers, networking devices, mobile phones, and any device that connects to a local network or internet.

macOS–The name of the operating system used by Apple Macintosh computers.

MB–*Megabyte*. A unit of information equal to 2^{20} bytes or approximately one million bytes.

MBA-IT–*Master of Business Administration, Information Technology specialization*. A graduate-level university program and degree designed for individuals seeking technology leadership positions.

MP–*Military Police*.

MX–*Mail Exchanger record*. Specifies the mail server responsible for accepting email messages on behalf of a domain name.

NIST–*National Institute of Standards and Technology* (USA-based). A physical sciences laboratory and non-regulatory agency of the United States Department of Commerce. Mission is to promote innovation and industrial competitiveness.

OMG–*Oh My God!*

OS–*Operating System*. The code running at the heart of computers.

PDF–*Portable Document Format*. A file format designed to include text, images, sound, video, and other media, in a manner independent of application software, hardware, and operating system.

PGP–*Pretty Good Privacy*. Cryptographic software.

PIN–*Personal Identification Number*. Typically used to validate electronic transactions.

PRISM–Code name for a program under which the United States National Security Agency collects internet communications from US internet companies.

QR Code–*Quick Response Code*. A machine-readable code consisting of an array of black and white squares. Typically used to store URLs and other information for reading by the camera on a smartphone.

RADIUS–*Remote Authentication Dial-In User Service*. A networking protocol providing centralized authentication, authorization, and accounting management for users connecting to a network service.

RAM–*Random Access Memory*. This is the area holding data as it is needed by the computer.

SIM–*Subscriber Identification Module*. A smart card inside a mobile phone, carrying an identification number unique to the owner, storing personal data, and preventing operation if removed.

S/MIME–*Secure/Multipurpose Internet Mail Extensions*. A standard for public key encryption and signing of MIME data.

SPF–*Sender Policy Framework*. An email validation system.

SOS–An international code signal of extreme distress.

SSD–*Solid-State Drive*. A storage device that uses integrated circuit assemblies to store data persistently.

SSL–*Secure Sockets Layer*. A digital encryption protocol.

TB–*Terabyte*. A unit of information equal to 2^{40} bytes or approximately one million million bytes.

TCP–*Transmission Control Protocol*. A set of rules for the delivery of data over a local network or internet.

TKIP–*Temporal Key Integrity Protocol*. An algorithm used to secure wireless networks.

TLS–*Transport Layer Security*. A digital encryption protocol.

Tor–*The Onion Router*. A highly secure web browser which uses both data encryption and random bouncing between multiple routers to help ensure security and privacy.

TXT–*Text record*. A type of resource record in the DNS used to provide the ability to associate arbitrary text with a host or other name, such as human readable information about a server, network, or data center.

US-CERT–*United States Computer Emergency Readiness Team*. Organized within the Department of Homeland Security's Cyber Security and Infrastructure Security Agency.

VPN–*Virtual Private Network*. A protocol/software that allows for end-to-end encrypted communication or data transmission over a local network or internet.

WEP–*Wired Equivalency Protocol*. An early encrypted wireless networking protocol.

WPA–*Wi-Fi Protected Access*. As of version 3, the current encrypted wireless networking protocol.

Index

www.ingramcontent.com/pod-product-compliance
Lightning Source LLC
Chambersburg PA
CBHW081046220326
41598CB00038B/6999